Developing Mathematical Literacy in the Context of the Fourth Industrial Revolution

Edgar Oliver Cardoso Espinosa
Instituto Politécnico Nacional, ESCA Santo Tomas, Mexico

A volume in the Advances in Educational Technologies and Instructional Design (AETID) Book Series

IGI Global
PUBLISHER of TIMELY KNOWLEDGE

Published in the United States of America by
IGI Global
Information Science Reference (an imprint of IGI Global)
701 E. Chocolate Avenue
Hershey PA, USA 17033
Tel: 717-533-8845
Fax: 717-533-8661
E-mail: cust@igi-global.com
Web site: http://www.igi-global.com

Library of Congress Cataloging-in-Publication Data

Names: Cardoso Espinosa, Edgar Oliver, 1973- editor.
Title: Developing mathematical literacy in the context of the Fourth Industrial Revolution / Edgar Oliver Cardoso Espinosa, editor.
Description: Hershey, PA : Business Science Reference, [2021] | Includes bibliographical references and index. | Summary: "This book provides empirical research or studies on the formation of mathematical abilities in the context of the Fourth Industrial Revolution regarding its development of both teaching and learning strategies, as well as the use of ICT in the development of this student knowledge"-- Provided by publisher.
Identifiers: LCCN 2020021647 (print) | LCCN 2020021648 (ebook) | ISBN 9781799838685 (hardcover) | ISBN 9781799864394 (paperback) | ISBN 9781799838692 (ebook)
Subjects: LCSH: Mathematics--Study and teaching. | Mathematics--Study and teaching--Technological innovations. | Mathematical readiness.
Classification: LCC QA11.2 .D48 2021 (print) | LCC QA11.2 (ebook) | DDC 510.71--dc23
LC record available at https://lccn.loc.gov/2020021647
LC ebook record available at https://lccn.loc.gov/2020021648

This book is published in the IGI Global book series Advances in Educational Technologies and Instructional Design (AETID) (ISSN: 2326-8905; eISSN: 2326-8913)

British Cataloguing in Publication Data
A Cataloguing in Publication record for this book is available from the British Library.

For electronic access to this publication, please contact: eresources@igi-global.com.

Advances in Educational Technologies and Instructional Design (AETID) Book Series

ISSN:2326-8905
EISSN:2326-8913

Editor-in-Chief: Lawrence A. Tomei Robert Morris University, USA

MISSION

Education has undergone, and continues to undergo, immense changes in the way it is enacted and distributed to both child and adult learners. In modern education, the traditional classroom learning experience has evolved to include technological resources and to provide online classroom opportunities to students of all ages regardless of their geographical locations. From distance education, Massive-Open-Online-Courses (MOOCs), and electronic tablets in the classroom, technology is now an integral part of learning and is also affecting the way educators communicate information to students.

The **Advances in Educational Technologies & Instructional Design (AETID) Book Series** explores new research and theories for facilitating learning and improving educational performance utilizing technological processes and resources. The series examines technologies that can be integrated into K-12 classrooms to improve skills and learning abilities in all subjects including STEM education and language learning. Additionally, it studies the emergence of fully online classrooms for young and adult learners alike, and the communication and accountability challenges that can arise. Trending topics that are covered include adaptive learning, game-based learning, virtual school environments, and social media effects. School administrators, educators, academicians, researchers, and students will find this series to be an excellent resource for the effective design and implementation of learning technologies in their classes.

COVERAGE

- Web 2.0 and Education
- Classroom Response Systems
- Educational Telecommunications
- Instructional Design Models
- Collaboration Tools
- Online Media in Classrooms
- Curriculum Development
- Bring-Your-Own-Device
- Adaptive Learning
- Digital Divide in Education

IGI Global is currently accepting manuscripts for publication within this series. To submit a proposal for a volume in this series, please contact our Acquisition Editors at Acquisitions@igi-global.com or visit: http://www.igi-global.com/publish/.

The Advances in Educational Technologies and Instructional Design (AETID) Book Series (ISSN 2326-8905) is published by IGI Global, 701 E. Chocolate Avenue, Hershey, PA 17033-1240, USA, www.igi-global.com. This series is composed of titles available for purchase individually; each title is edited to be contextually exclusive from any other title within the series. For pricing and ordering information please visit http://www.igi-global.com/book-series/advances-educational technologies-instructional-design/73678. Postmaster: Send all address changes to above address.

Titles in this Series

For a list of additional titles in this series, please visit:
http://www.igi-global.com/book-series/advances-educational-technologies-instructional-design/73678

Cases on Active Blended Learning in Higher Education
Brenda Cecilia Padilla Rodríguez (Universidad Autonoma de Nuevo Leon, Mexico) and Alejandro Armellini (University of Northampton, UK)
Information Science Reference • © 2021 • 375pp • H/C (ISBN: 9781799878568) • US $195.00

Applying Universal Design for Learning Across Disciplines Case Studies on Implementation
Frederic Fovet (Royal Roads University, Canada)
Information Science Reference • © 2021 • 335pp • H/C (ISBN: 9781799871064) • US $195.00

Advancing Online Course Design and Pedagogy for the 21st Century Learning Environment
Daniel Chatham (Middlebury Institute of International Studies at Monterey, USA)
Information Science Reference • © 2021 • 382pp • H/C (ISBN: 9781799855989) • US $195.00

Affordances and Constraints of Mobile Phone Use in English Language Arts Classrooms
Clarice M. Moran (Appalachian State University, USA)
Information Science Reference • © 2021 • 229pp • H/C (ISBN: 9781799858058) • US $175.00

Examining an Operational Approach to Teaching Probability
Alessio Drivet (Geogebra Institute of Turin, Italy)
Information Science Reference • © 2021 • 369pp • H/C (ISBN: 9781799838715) • US $195.00

For an entire list of titles in this series, please visit:
http://www.igi-global.com/book-series/advances-educational-technologies-instructional-design/73678

IGI Global
PUBLISHER of TIMELY KNOWLEDGE

701 East Chocolate Avenue, Hershey, PA 17033, USA
Tel: 717-533-8845 x100 • Fax: 717-533-8661
E-Mail: cust@igi-global.com • www.igi-global.com

Editorial Advisory Board

Table of Contents

Detailed Table of Contents

The 21st century has witnessed the development of the fourth industrial revolution that means cutting-edge technologies including artificial intelligence and the internet of things. The special quality of this epochal revolution lies in a rapid acceleration of the speed of change, never seen before. This is happening at the most challenging time ever faced by our species, due to existential threats. This scenario begs big questions to professionals in the field of mathematics education. Can mathematics education contribute to build a better world? The chapter reports on a participatory research conducted by a researcher in mathematics education together with 11 mathematics teachers. Therefore, the research benefits from a double point of view, coming from different cultures so gaining a double verisimilitude, which makes the results acceptable for all the professionals involved.

This chapter aims to provide an in-service teaching model, which is named as a "reflective teachers-as-designers professional development model" based on the literature, designed to give teachers an understanding of teaching that focuses on developing real-life contexts and developing global competencies. First, the conceptual framework for the learning of teachers and their capacity as designers

Living in the 21st century brings new challenges for the educational system, especially that which concerns education in mathematics, statistics, and technologies. In this chapter, recent problems of teaching statistics in schools in Bulgaria are discussed. The present state of the mathematics curriculum is presented, namely in the section "Probabilities and statistics"; some existing problems are discussed, and recommendations are made for further improvement of the teaching material. The objective of the study is to provide ground for increasing the statistical literacy of the younger generation through a more comprehensive and more up-to-date education in school.

The objective of the chapter is the elaboration of an intervention proposal based on the STEAM Integrator Project that from mathematics addresses the issues related to the 2030 Agenda proposed by the United Nations. The central element of the proposal is to consider the context and interest of the student for the selection of a topic related to science to explain it, argue it, and describe it in order to model it from different disciplines of the physical-mathematical area in the Upper Middle Level of the Instituto Politécnico Nacional.

Mathematics is a fundamental human activity – a way of making sense of the world. Children possess a natural curiosity and interest in mathematics and come to school with an understanding of mathematical concepts and problem-solving strategies that they have discovered through explorations of the world around them. The purpose of this study is to find out how mathematics is being taught and learnt among primary one pupils in basic schools in Ghana. A qualitative approach was employed to provide answers to the research questions raised. It was found that teachers employed variety of procedures for teaching mathematics. However, there were some problems encountered in the teaching and learning of mathematics.

Moving into the fourth industrial revolution and the rapid digital transformation, there is a huge volume of data to be managed in each industry. Industrial simulations commonly produce data including the inputs and outputs of linear systems with several million unknowns. Solving linear systems is one of the fundamental problems in scientific computing, and it requires significant system resources. Determining a suitable method to solve linear systems can be a challenging task, since there is not a certain knowledge about which method is the most suitable for different numerical problems. In this study, the authors demonstrate how machine learning (ML) approach can be used in selecting solvers for linear systems. The chapter includes frequently used ML methods from literature and explain the usage of them to select optimal solvers and preconditioners.

The objective of the chapter is the elaboration of a methodological proposal for project-oriented learning at the postgraduate level in order to develop students' mathematical competencies. The methodology of project oriented learning has as its central element the elaboration of a project based on a problem to propose a solution through a research processs. Thus, it recognizes the importance of integrating both participation and collaborative work in the training practices of graduate students to achieve quality training in their learning, which allows a transfer of skills in their work performance. In the same way, the integration of the digital competence, communication, interpersonal, and critical thinking is incorporated.

Preface

The term *revolution* refers to the presence of an abrupt and radical change. Thus, revolutions have been generated throughout history when new technologies and novel ways of interacting with the world promote transformations in economic systems and social structures. In the same way, in each of these revolutions, the economic, social and technological models are obsolete and those of the future are not yet fully constructed or socially accepted as true. In the last two centuries there have been three industrial revolutions in which the main sources of energy changed, the type of most important industrial activities with their geographical location, as well as the means of communication used both to displace people and goods (Blanco, Castro, Gayoso and Santana, 2019).

According to Daemmrich (2017), the First Industrial Revolution emerged at the end of the 18th century in the United Kingdom with the introduction of steam power, the mechanization of agriculture, manufacturing and transport. The Second Industrial Revolution was consolidated in the United States of America at the beginning of the 20th century, where the disruptive technology was electrical energy, which allowed serial production (Hintz, 2011; MacLeod, 2016). The Third Industrial Revolution was triggered since the middle of the last century with the invention of semiconductors, digital platforms and the rise of personal computers (Hermann, Pentek and Otto, 2016).

Unlike the previous three, the Fourth Industrial Revolution is not defined by the emergence of a specific disruptive technology, but by the convergence of various digital, physical and biological technologies, such as artificial intelligence, augmented intelligence, robotics, 3D printing, *cloud computing*, *big data*, the Internet of things or nanotechnology (Ynzunza, Izar, Bocarando, Aguilar and Larios, 2017).

Hence, the Fourth Industrial Revolution impacts on companies, as well as on the ways of working between individuals, which implies the development of new forms of production, organization, distribution and exchange between the various agents and sectors of a country (Schwab, 2015). In this sense, companies have incorporated new processes related to global value chains in which the use of various technologies is constant. In addition, in the future, organizations will establish global networks that will incorporate machinery, storage systems and production strategies with cyber-physical systems (Kagermann, Wahlster and Helbig, 2013).

So, other fundamental features such as speed at an exponential rate due to technological innovations; breadth and depth based on the use of multiple technologies that impact the economy, business, society and people, as well as an impact on the systems among its various components at the country, sectorial, organizational and intersectorial level arise (Schwab, 2017).

Based on the above, in the Fourth Industrial Revolution, a world of continuous changes is envisioned, where science and technology are at the base of the economic competitiveness of countries, highlighting the strategic role of innovation, which implies meeting a series of requirements to compete successfully in an increasingly globalized economy, including sustained investment in both education and research. In the same way, both organizations and people are required to acquire new qualities to adapt to the accelerated changes in this new context, which implies continuous or permanent training (Redecker and Punie, 2013).

Therefore, the Fourth Industrial Revolution is more than a change driven by technology because it is a phenomenon oriented towards innovation to positively impact the central sectors of a country such as education, health and business. Specifically in the first, with the previous industrial revolutions, the focus changed because in the Fourth Industrial Revolution, technologies expanded in various ways the training modalities between face-to-face and virtual, thus requiring a redefinition of their substantive functions considering new modes of curriculum and teaching, and consequently, a transformation of the various forms of learning, which has led to the relevance of designing alternative curricula (Jules, 2017).

Accordingly, educational institutions have to reinvent themselves, evolve and innovate towards the construction of flexible learning experiences in the student body in relation to what, where, when and how. In addition, they will be relevant because they will empower the elimination of barriers due to the fact that they will allow access to the education without borders of time or space, thus increasing collaboration between people regardless of their geographic location, and generating information exchange networks (Castaño, Redecker, Vuorikari and Punie, 2014).

On the other hand, Barrios and Fajardo (2017) mention the generation of a digital space, which fosters a virtual society scenario where meeting points are established, friendship ties are consolidated and communication mechanisms are strengthened, so that it constitutes the center of information, but at the same time, a learning agora that is useful because it optimizes the search processes as well as determines that education has to transform itself in the face of this new scenario called the Fourth Industrial Revolution.

Meanwhile, Leymonié (2015) establishes a new perspective of educational training oriented towards the management of capacities in relation to their development, links with the labor sector and decision-making, so that human talent is the fundamental element that enables both the design and the implementation of innovative strategies in organizations that allow solving the various situations and changing conditions in the markets, as well as promoting not only economic but comprehensive growth. To achieve this, it is expected that educational training is focused on the development of basic and specific competencies such as the ability to work in a team, the development of autonomous thinking and the willingness to lifelong learning.

Consequently, the Fourth Industrial Revolution is having an impact on educational organizations with respect to the academic training they provide, which is why new competencies associated with the transfer of knowledge are required because the job profiles of the future will require human talent in multidisciplinary skills that allow companies to create and improve the operation of machinery and technological devices under the scheme of an integrated system (Pejic, Bertoncel, Mesko and Drstic 2020).

In this situation, both the World Economic Forum (WEF, 2015) and the McKinsey Global Institute (MGI, 2017a) indicated that the basic competencies to be developed are skills focused on science, technology, engineering and mathematics (STEM approach), as well as focusing on promoting creativity and critical thinking. This is due to the fact that human capital does not only require the capacity for digitization, but also specific knowledge both to solve problems related to technology in an optimal way and to make decisions based on the data obtained (Carrillo, Gomis, De los Santos, Covarrubias and Matus, 2020).

Given this panorama, the training of mathematical learning is important because it is oriented towards the development of a set of capacities in order to solve situations of daily and professional life, so it focuses on the acquisition of different ways of representing the information through models, constructions and graphs to determine the best decision making. In this sense, it includes the domain of representations, measures and quantitative structures to carry out the interpretation of the results in specific contexts, so that mathematics retains its main place within the academic training of students in the 21st century since basic education to higher level (OECD, 2018). This is because it has been identified that for a society to promote innovation, the use of mathematical information is an essential condition that allows the development of creativity and information analysis (Pfeiffer, 2017).

Similarly, in the context of the Fourth Industrial Revolution, any information is presented in different ways: graphic, numerical, and geometric coupled with the fact that it is accompanied by statistical arguments. Likewise, a mathematical training allows the individual a set of tools of universal value in which he develops his understanding and reasoning that allows him to make decisions both on a personal and professional level. In addition, the increasing influence of Information and Communication Technologies (ICTs) at all levels of the educational system of a country where they have become mediating tools for the formation of mathematical abilities is unquestionable (Yang, 2017).

Therefore, mathematical capacities are the set of abilities, skills and attitudes to develop and apply reasoning in order to solve situations of daily and professional life, so it is oriented towards the acquisition of using the different ways of representing the information in the form of models, constructions and graphs to determine better decision-making (Zoldos, 2015). In this sense, it includes mastering the handling of terminology, measurements and formal structures to carry out the interpretation of procedures and representations of a quantitative nature on personal and professional situations. Also, it includes a positive attitude towards its employment aimed at recognizing its usefulness and validity in the professional field (Cardoso, 2019).

In the same way, it has been recognized that favoring mathematical skills allows the development of other transversal skills such as commitment, responsibility, decision-making, interpersonal communication, flexibility, time management, leadership, conflict resolution and teamwork. According to Deming (2015), the development of the above competencies in people generates a quality educational training because they have greater employment opportunities. Similarly, another element to integrate is the diversity of technological tools in terms of their access, assisted learning, interventions or mediations by teachers and formative assessment (Bettinger, Fox, Loeb and Taylor, 2017).

Consequently, it is necessary for the human capital of any organization to acquire a quality mathematical training that allows it to master a series of capacities to carry out tasks efficiently and effectively in the various areas of life, where it recognizes the existence of the information represented in a quantitative way, so its organization, analysis and presentation is required through digital tools (Galindo, Ruíz and Ruíz, 2017).

To achieve this, education in the context of the Fourth Industrial Revolution needs to be transformed to assume a proactive role in the development of innovation in the country, becoming the main factor in its comprehensive development in which mathematics is one of the main subjects to be considered in the educational journey of citizens. Given this panorama, it is important that teachers of any educational level reorient their teaching strategies and their role as educators (Flores, 2017).

Hence, the panorama described above implies a great responsibility for the educational system of a country, which is why it is a relevant topic to investigate because its development includes both the formation of knowledge, skills and attitudes of an epistemological, social and psychopedagogical type in the students as well as the design, implementation and evaluation of teaching strategies at the various educational levels. Based on the above, the objective of the publication is to provide research or empirical studies on the formation of mathematical abilities in the context of the Fourth Industrial Revolution oriented towards the formulation, development and evaluation of strategies for both teaching and learning in combination with the use of ICTs.

It should be noted that the process carried out for the preparation of this publication included the following phases by the editor: 1) Sending a book proposal describing the coverage and importance of the subject for evaluation by the Advisory Board of the IGI Global Editorial ; 2) Selection of the evaluation committee; 3) Invitation at the international level of the chapter proposals; 4) Performing the double blind review and ruling for each proposal by at least two evaluators; 5) Mailing to the authors about the reviews of each evaluator for compliance; 6) Compilation and organization of the chapters that met the evaluation and format criteria and 7) Preparation of the final work.

Finally, it is important to note that the evaluation of each chapter that integrates this work meets the following criteria:

- Originality and meaning of the contribution consistent with the theme of the book.
- International relevance of the chapter theme and its contributions.
- Coverage of existing prior literature.
- Proper use of the methodology used with the respective foundation.
- Establishment of the conclusions and previous studies derived from a research process.
- Compliance with the formal structure for a chapter.

ORGANIZATION OF THE BOOK

The book is organized in nine chapters. Below is a brief description of each of the chapters:

Chapter 1 is oriented reports on a participatory research conducted by a researcher in Mathematics Education together with eleven Mathematics teachers. Therefore, the research benefits from a double point of view, coming from different cultures so gaining a double verisimilitude, which makes the results acceptable for all the professionals, involved. The 21st century has witnessed the development of the Fourth industrial revolution that means cutting-edge technologies including Artificial Intelligence and the Internet of Things. The special quality of this epochal revolution lies in a rapid acceleration of the speed of change, never seen before. This is happening at the most challenging time ever faced by our species, due to existential threats. This scenario begs big questions to professionals in the field of Mathematics education. Can Mathematics Education contribute to build a better world?

Chapter 2 aims to provide an in-service teaching model, which is named as a "reflective teachers-as-designers professional development model" based on the literature, designed to give teachers an understanding of teaching that focuses on developing real-life contexts and developing global competencies. First, the conceptual framework for the learning of teachers and their capacity as designers is discussed. Then, content components and process components of the model are presented by providing the rationale behind each element.

Chapter 3 addresses the topic of Big Data in order to determine its importance in the mathematical training of students. The objective of the chapter is to determine the importance of the incorporation of big data in mathematical training according to the context of the fourth industrial revolution through a documentary research that will be of interest to teachers who are dedicated to mathematics training of students in this area of knowledge.

Chapter 4 deals the study investigated students' attitude in MST education for digitally-based society in the South-West region, Nigeria. Findings revealed that the students in the South-West geopolitical region had a poor attitude towards Mathematics, Chemistry, Physics and Computer Science, and there was a poor academic performance in MST education. Recommendations on periodical workshops and training of teachers on the innovativeness in the teaching were made.

Chapter 5 proposes recent problems of teaching statistics in schools in Bulgaria are discussed. The present state of the mathematics curriculum is presented, namely in the section "Probabilities and statistics"; some existing problems are discussed, and recommendations are made for further improvement of the teaching material. The objective of the study is to provide ground for increasing the statistical literacy of the younger generation through a more comprehensive and more up-to-date education in school.

Chapter 6 is geared toward the STEAM project. The objective of the chapter is the elaboration of an intervention proposal based on the STEAM Integrator Project that from mathematics addresses the issues related to the 2030 Agenda proposed by the United Nations. The central element of the proposal is to consider the context and interest of the student for the selection of a topic related to science to explain it, argue it and describe it in order to model it from different disciplines of the physical-mathematical area in the Upper Middle Level of the Instituto Politécnico Nacional.

Chapter 7 aims to find out how mathematics is being taught and learnt among Primary one pupils in basic schools in Ghana. A qualitative approach was employed to provide answers to the research questions raised. It was found that teachers employed variety of procedures for teaching mathematics. However, there were some problems encountered in the teaching and learning of mathematics. Therefore, Mathematics is a fundamental human activity – a way of making sense of the world. Children possess a natural curiosity and interest in mathematics and come to school with an understanding of mathematical concepts and problem-solving strategies that they have discovered through explorations of the world around them.

Chapter 8 is oriented towards a proposal to use the machine learning (ML) approach can be used in selecting solvers for linear systems. Our paper includes frequently used ML methods from literature and explain the usage of them to select optimal solvers and preconditions'. This is based on the fact as moving into the Fourth Industrial Revolution and the rapid digital transformation, there is a huge volume of data to be managed in each industry. Industrial simulations commonly produce data including the inputs and outputs of linear systems with several million unknowns. Solving linear systems is one of the fundamental problems in scientific computing, and it requires significant system resources. Determining a suitable method to solve linear systems can be a challenging task, since there is not a certain knowledge about which method is the most suitable for different numerical problems.

Chapter 9 has the objective of developing a methodological proposal for Project Oriented Learning at the postgraduate level in order to develop students' mathematical competencies. The methodology of Project Oriented Learning has as its central element the elaboration of a project based on a problem to propose a solution through a research process. Thus, it recognizes the importance of integrating both participation and collaborative work in the training practices of graduate students to achieve quality training in their learning, which allows a transfer of skills in their work performance. In the same way, the integration of the digital competence, communication, interpersonal and critical thinking is incorporated.

FINAL THOUGHTS

This book provides information on mathematics training in educational programs for the Fourth Industrial Revolution. This is an innovative and important topic that deserves reflection and this book is an excellent opportunity to do so, increasing understanding on this subject, providing relevant academic work, empirical research findings and an overview of this relevant field of study.

It is a timely research publication that presents the latest academic trends, problems and implications for quality mathematical training in the countries, including a wide coverage of relevant topics such as the incorporation of didactic methodologies, incorporation and use of technologies for the teaching of mathematics, interesting studies on students' attitudes towards mathematics, as well as the teaching skills to teach this subject.

Finally, this work establishes the importance of implementing changes in educational programs with an innovative approach that promote mathematical competencies in students that allow them not only to apply procedures but also to develop critical thinking, reasoning, argumentation and communication. Likewise, it provides teaching strategies for its implementation, which will have an impact on quality training for students. Therefore, it offers cutting-edge information on mathematical training in the context of the Fourth Industrial Revolution.

ACKNOWLEDGMENT

The editor is grateful to the Instituto Politécnico Nacional (IPN) of Mexico for the facilities granted to carry out this book from the authorization of the two-year research project entitled: Evaluation of the impact of the academic training of postgraduate degrees in administration with numbers of registration SIP20195564 and SIP20200768.

REFERENCES

Barrios, A., & Fajardo, G. (2017). El ecosistema educativo universitario impactado por las TIC. *Anagramas*, *15*(30), 101–120. doi:10.22395/angr.v15n30a5

Bettinger, E., Fox, L., Loeb, S., & Taylor, E. (2017). Virtual Classrooms: How Online College Courses Affect Student Success. *The American Economic Review*, *107*(9), 2855–2875. doi:10.1257/aer.20151193

Blanco, F., Castro, J., Gayoso, R., & Santana, W. (2019). *Las claves de la Cuarta Revolución Industrial*. Libros de Cabecera.

Cardoso, E. (2019). Las actitudes hacia las matemáticas de estudiantes de formación inicial de profesorado en México. *Journal of Psychology and Behavioral Sciences of the Academic Unit of Social and Legal Sciences*, *10*(1), 87–103. doi:10.29059/rpcc.20190602-83

Carrillo, J., Gomis, R., De los Santos, S., Covarrubias, L., & Matus, M. (2020). ¿Podrán transitar los ingenieros a la Industria 4.0? Análisis industrial en Baja California. *Entreciencias: Dialogues in the Knowledge Society*, *8*(22), 1–22. doi:10.22201/enesl.20078064e.2020.22.76089

Castaño, J., Redecker, C., Vuorikari, R., & Punie, Y. (2014). Open Education 2030: Planning the future of adult learning in Europe. *Open Learning*, *28*(3), 171–186. doi:10.1080/02680513.2013.871199

Daemmrich, A. (2017). Invention, Innovation Systems, and the Fourth Industrial Revolution. *Technology and Innovation*, *18*(4), 257–265. doi:10.21300/18.4.2017.257

Deming, D. (2015). The growing importance of social skills in the labor market. *NBER Working Paper*, 21473. Retrieved from https://www.nber.org/papers/w21473

Flores, W. (2017). Incorporación de tecnologías en la enseñanza de las matemáticas: Actitudes del estudiantado universitario. *Pedagogical Horizons, 19*(1), 21-30. Retrieved from https://revistas.iberoamericana.edu.co/index.php/rhpedagogicos/article/view/1101/1008

Galindo, F., Ruíz, S., & Ruíz, F. (2017). Competencias digitales ante la irrupción de la Cuarta Revolución Industrial. *Journal Estudos em Comunicacao, 25*(1), 1–11. http://ojs.labcom-ifp.ubi.pt/index.php/ec/article/view/277/144

Hermann, M., Pentek, T., & Otto, B. (2016). Design Principles for Industrie 4.0 Scenarios. *Proceedings of 49th Hawaii International Conference on System Sciences HICSS, 5-8*, 3928-3937. 10.1109/HICSS.2016.488

Hintz, J. (2011). Peer educators responding to an institucional challenge: campus student services. *Emerging Issues and Practices in Peer Education, 133*, 87-95. Retrieved from https://doi doi:lorg/10.1002s.387

Jules, T. (2017). *Public Policy and Governance. The Global Educational Policy Environment in the Fourth Industrial Revolution: Gate, Regulated and Governed.* Emerald Group Publishing Limited.

Kagermann, H., Wahlster, W., & Helbig, J. (2013). *Securing the future of German manufacturing industry. Recommendations for implementing the strategic initiative INDUSTRIE 4.0. Final report of the Industrie 4.0 Working Group*. Acatech.

Leymonié, J. (2015). Nativos e inmigrantes digitales: ¿cómo aprendemos y enseñamos? *Dixit, 12*, 10–19. doi:10.22235/d.v0i12.292

MacLeod, A. (2016). Prospection, well-being and memory. *Memory Studies, 9*(3), 266–274. doi:10.1177/1750698016645233

McKinsey Global Institute. (2017a). *What's now and next in analytics, AI, and automation.* Retrieved from https://www.mckinsey.com/featured-insights/digital-disruption/whats-now-and-next-in-analytics-ai-and-automation#

OECD. (2018). *Assessment and Analytical Framework: Mathematics, Reading, Science, Problem Solving and Financial Literacy (PISA, 2018)*. OECD Publishing. doi:10.1787/1996377

Pejic, M., Bertoncel, T., Mesko, M. & Drstic, Z. (2020). Text mining of industry 4.0 job advertisements. International *Journal of Information Management, 50*, 416-431. .ijinfomgt.2019.07.014 doi:10.1016/j

Pfeiffer, S. (2017). The Visión of "Industrie 4.0" in the Making – a Case of Future Told, Tamed, and Traded. *NanoEthics*, *11*(1), 107–121. doi:10.100711569-016-0280-3 PMID:28435474

Redecker, C., & Punie, Y. (2013). The future of learning 2025: Developing a vision for change. *Future Learning*, *1*, 3–17.

Schwab, K. (2015). The Fourth Industrial Revolution. What It Means and how to Respond. *Foreing Affairs*. Retrieved from https://www.foreignaffairs.com/articles/2015-12-12/fourth-industrial-revolution

Schwab, K. (2017). *The Fourth Industrial Revolution*. World Economic Forum.

World Economic Forum. (2015). *New Vision for Education. Unlocking the Potential of Technology*. Retrieved from https://widgets.weforum.org/nve-2015/

Yang, L. (2017). Industry 4.0: A survey on technologies, applications and open research issues. *Journal of Industrial Information Integration, 6,* 1-10. doi:10.1016/j.jii.2017.04.005

Ynzunza, C., Izar, L., Bocarando, C., Aguilar, P., & Larios, O. (2017). El Entorno de la Industria 4.0: Implicaciones y Perspectivas Futuras. *Technological Awareness*, (54). Retrieved from https://www.redalyc.org/articulo.oa?id=94454631006

Zsoldos, I. (2015). Changing pre-service primary-school teachers' attitude towards Mathematics by collaborative problem solving. *Procedia: Social and Behavioral Sciences*, *186*, 174–182. doi:10.1016/j.sbspro.2015.04.100

Acknowledgment

The editor would like to acknowledge the help of all the people involved in this project and, more specifically, to the authors and reviewers that took part in the review process. Without their support, this book would not have become a reality.

First, the editor would like to thank each one of the authors for their contributions. Our sincere gratitude goes to the chapter's authors who contributed their time and expertise to this book.

Second, the editor wishes to acknowledge the valuable contributions of the reviewers regarding the improvement of quality, coherence, and content presentation of chapters. Most of the authors also served as referees; we highly appreciate their double task.

Third, the editor is grateful to the Instituto Politécnico Nacional (IPN), Mexico for the facilities granted to carry out this book from the authorization of the two-year research project entitled: Assessment of the impact of the academic training of postgraduate degrees in administration with numbers of registration SIP20195564 and SIP20200768.

Edgar Oliver Cardoso Espinosa
Instituto Politécnico Nacional, ESCA Santo Tomás, Mexico

Chapter 1
Professional Development of Mathematics Teachers in the Age of I4.0:
Can Mathematics Education Contribute to Build a Better World?

Maria Guida
Istituto Nazionale di Documentazione, Innovazione e Ricerca Educativa, Italy

ABSTRACT

The 21st century has witnessed the development of the fourth industrial revolution that means cutting-edge technologies including artificial intelligence and the internet of things. The special quality of this epochal revolution lies in a rapid acceleration of the speed of change, never seen before. This is happening at the most challenging time ever faced by our species, due to existential threats. This scenario begs big questions to professionals in the field of mathematics education. Can mathematics education contribute to build a better world? The chapter reports on a participatory research conducted by a researcher in mathematics education together with 11 mathematics teachers. Therefore, the research benefits from a double point of view, coming from different cultures so gaining a double verisimilitude, which makes the results acceptable for all the professionals involved.

DOI: 10.4018/978-1-7998-3868-5.ch001

INTRODUCTION

In the 21st century, we have witnessed the development of the Internet of Things that, connecting to production techniques, has allowed systems to collect and share information, analyze it and use it to guide data driver decision. In the meantime, cutting-edge technologies including artificial intelligence, robotics, additive manufacturing, machine learning and other cognitive technologies, advanced materials and augmented reality have become established and widespread. This epochal change is known as the Fourth Industrial Revolution or Industry 4.0.

The special quality of this digital revolution lies in a rapid acceleration of the speed of change, never seen before. First consequence on education is that progress is now so quick that the obsolescence of knowledge is shorter and shorter which means that by the time our students apply what they know, it is already outdated.

In addition Industry 4.0 is a technological revolution that takes place in a world that is already complex itself and in a century which is the most challenging that our species has ever faced. Science tells us that during this century we will also face existential threats in the form of climate change, rapid global population growth, ageing populations, growing inequality, antibiotic resistance and ideology. The great challenges of the present and future involve the fate of our planet and put it in relation to the establishment of a new way of acting in humanity.

The depicted scenario begs big questions to professionals in the field of education in general and mathematical education in particular. What is worth learning in mathematics in the context of the fourth industrial revolution? Which topics are worth teaching to develop a broader intelligence that prepares students to face a future that we do not yet know? Which is the topic selection criterion, given that in a labor market rapidly changing, whatever topic goes quickly out to date and it is difficult to prepare future workers for new jobs that nobody knows yet? How can mathematical education contribute to creating aware citizens in an increasingly complex and challenging world? Which characteristics and which skills does a citizen need to be able to actively assume his place in this society? What do we need to do for the next generation of young people to optimize their potential?

In this uncertain time, looking through these big questions, a few things look clear.

The first one is that the current/old model of schooling is no longer working. The technological revolution will place significant challenges over many years on employment, income security and social agency for large populations across social and cultural groups. At odds yet nothing in our formal education systems today is designed to meet these human challenges of the 21st century. Despite the many speeches, the school generally still follows the factory model. The classrooms themselves, ordered by rows of individual desks, are reminiscent of the English textile industries of the mid-nineteenth century, in which the machines were positioned on small-aligned tables, behind which the workers sat. People still are generally considered educated when they possess a large collection of facts and procedures so the core activity of the classroom is still instruction and standardized test made by the teacher who requires conformism from students that assimilate notions passively. This was called "standard model" which goals were to ensure standardization where all students memorized and master the same core curriculum. The analogies between school and the industrial-age factory were made explicit (Callahan, 1962), and this structural alignment facilitated the ease of transition from school student to factory worker. However, this approach appears unsuitable for the 4.0 factories that will increasingly require a creative, adaptable and flexible workforce, capable of facing a technological progress in continuous evolution and acceleration. So nowadays Industry is 4.0, school instead is still 2.0 or sometimes 3.0 with the adoption of new technological tools unless they just dress up the usual old practices, like a sprinkling of sugar on the bitter pill, which unfortunately sometimes happens.

The second thing is that teachers can be the lever of change, because they can contribute to change the school from inside on a daily basis. In order to do this they should be offered professional development opportunities. Nevertheless, if for students the transmission method has proved ineffective, we cannot think of offering teachers courses based on the transmission of theoretical contents, neither in presence nor in fad. Rather, they need opportunities to study, discuss and reflect on their teaching practices in the context of action, building a community of practices with their peers.

Given the strong evidence for the failure and inadequacy of the old industrial model of schooling, Valerie Hannon (2017) argues that the purpose of education nowadays is about learning how to thrive in a transforming world. Children and young people need to be allowed to experience and do at school whatever allow them to learn to live in new and better ways, taking charge of the future so they can shape it and deal with its challenges and not just to develop a set of abstract skills,. Thinking about the future and how the world has already changed, and will change exponentially faster, should direct us towards new purposes for education. Some of these new purposes will be collective, rather than just individual or national. Through the experience of the Covid-19 pandemic, we learn that those who confuse social distancing with isolation are wrong because this is a global challenge and will only be won with everyone's collaboration. The virus recognizes no borders between states and has exacerbated inequalities and situations of poverty that we have to face collaboratively, at a global level, on a world scale. Risks to economic stability and social cohesion continue to intensify.

Effective worker figures, as mentioned above, must be supported by a diverse set of skills that the school should equip them with. However, school is not just an appendage to the business world! Moreover, it does not just teach how to be more competitive to get a more profitable job.

The school must help train people who build a better world with more well-being and less inequality. This will improve everyone's quality of life. Developing crucial global citizenship means attitude to innovation, creativity, technological skills but also interpersonal skills like emotional intelligence and interpersonal communication both for responding to the labor market request and in order to make a better World.

However, established that, what can mathematics education do? First, it is necessary to acquire a broader view of mathematics teaching and of the role that mathematical education plays in society.

In order to train citizens competent for life, education must be seen as a permanent effort to develop creativity, curiosity, critical thinking, planning thinking, a global mentality, that is, knowing how to look at phenomena in a systematic way, seizing relationships and interdependencies between them.

It can be done through mathematics teachers setting aside three old standards: the tayloristic and transmissive model of schooling, the coverage syndrome and the emphasis on procedural skills.

As a matter of fact, making effort on students' procedural fluency is a waste of energy because they tend to forget so at best, a short-term result will be obtained. At the opposite, a deep conceptual understanding equips students with a mental framework to be considered a long-term result.

It is very important indeed to identify the "Throughlines", that are the core topics of mathematics, including specific methods and languages of the disciplinary domain. Throughlines help teachers focus on essential elements of the curriculum because, according to Howard Gardner (1983), "the greatest enemy of understanding is coverage. As long as you are determined to cover everything, you actually ensure that most kids are not going to understand. You've got to take enough time to get kids deeply involved in something so they can think about it in lots of different ways and apply it—not just at school but at home and on the street and so on.".

Last, but not least, not just one aspect but the whole model of schooling must be changed. Learning environments, global perspective, cross-discipline questions would be key elements and the transmittive methods should be replace by the more effective ones such as project based collaborative learning with a focus on personalization, inclusivity and the perspective of lifelong continuous learning. Learning environments should be created in order for students to make experiences and mistakes and to work together, not isolated, to tackle complex problems together and to propose solutions, enhancing the informal dimension of learning, developing soft skills that are indispensable for an approach to changing and complex scenarios. Thus mathematics importance emerges clearly when they use it as a lens to understand the great problems of today's world. The nature of the task should be changed if we want students to develop a "productive disposition— [the] habitual inclination to see mathematics as sensible, useful, and worthwhile, coupled with a belief in diligence and one's own efficacy." (NRC, 2001)

This is not achieved if we engage students only in the solution of exercises that focus attention on the product, the solution that must be exactly what is written on the book. The exercises are based on necessary and sufficient knowledge, are the application of someone else's discovery, reproduce known patterns, require the application of previously memorized techniques. On the contrary, in problems, knowledge is necessary but not sufficient, they require making a discovery that is the result of creativity, intuition, reasoning, structuring. Attention is paid to the solution process, not to the product.

Of course, the renewal of teaching in mathematics is a slow process that actualizes gradually and can only take place involving teachers in an extensive professional development action over time, with a teaching-research design, in which class practice is inseparable from educational theories for improving learning in the context of action.

The chapter reports on a research project conducted with a participatory approach by a researcher in mathematics education together with eleven mathematics teachers, the entire math department of a technical-vocational school in the South of Italy, assuming in the project the role of co-researchers. The work therefore benefits from a double point of view, coming from different cultures so gaining a double verisimilitude (Dubet, 1994) which makes the results acceptable for all the professional involved.

The chapter aims to affirm that some problematic aspects of mathematics teaching can be addressed thanks to the Making Learning and Thinking Visible framework. In particular the discussion protocols developed in the framework and the planning and experimentation in the classroom of structured teaching units through its TRs, offer teachers the opportunity for professional development from which transformative learning derives which changes the practices and also the culture of the school as a whole.

BACKGROUND

The Need to Unlearn

The futurist Alvin Toffler (1970) is quoted everywhere with his sentence: "The illiterate of the 21st century will not be those who cannot read and write, but those who cannot learn, unlearn, and relearn". Moreover, Anshul Sonak, Intel's director of education for the Asian area, says accordingly that today the new paradigm should be: "learn, fail, unlearn, re-learn". On the same wavelength, a team of Harvard researchers has studied the individual and organizational mechanisms that make change difficult, claiming that in order to change it is necessary to abandon some patterns and routines, to recover intentionality and reflexivity. They created a framework with a provocative name "Unlearning to learn", (Ritchhart, 2015) which suggests openness to creativity, to welcome unusual solutions, new possibilities and the need to cancel practices, habits and methods acquired over time, in order to adopt new ones. Unlearning is not about forgetting. It's about the ability to choose an alternative mental model or paradigm. When we learn, we add new skills or knowledge to what we already know. When we unlearn, we step outside the mental model in order to choose a different one. Toffler was emphasizing that in times of rapid change in an uncertain future the most valuable skill would be learning how to learn rather than simply reciting a set of facts and procedures. Designing learning experiences that encourage learners to make things by collaboratively solving interesting real world challenges will be the key to thriving in this century. We will succeed by working alongside our machines rather than competing with them, by programming them rather than being programmed by them.

Mathematical Literacy and Empowerment

Within the framework of the OECD Programme for International Student Assessment (PISA, 2006), mathematical literacy is defined as "an individual's capacity to identify and understand the role that mathematics plays in the world, to make well-founded judgements and to use and engage with mathematics in ways that meet the needs of that individual's life as a constructive, concerned and reflective citizen. Mathematical literacy is concerned with the ability of students to analyze, reason, and communicate ideas effectively as they pose, formulate, solve, and interpret solutions to mathematical problems in a variety of situations" (OECD, 2009).

Mathematical literacy focusing on citizenship should refer to the aim of critically evaluating aspects of the surrounding culture - a culture that is more or less colonized by practices that involve mathematics. Thus, the ability to understand and to evaluate these practices should form a component of mathematical literacy (Jablonka, 2003).

Math competence has a threefold dimension dealing with knowledge, ability to solve problems and reasoning. As to the first dimension, knowledge, it refers to instrumental mathematical skills concerning holding and use of mathematical notions, their representations, their properties, related operational techniques and the relationships between the notions themselves. As to the second dimension, problem solving ability, it refers to the skills related to the information needed to solve a problem, constructing or identifying the appropriate reasoning, with attention to the representation tools used and account for it, checking this reasoning and the results that follow in relation to the problem considered. The third dimension, argumentation, refers to be able to ascertain the validity of a statement or a proceeding or a reasoning, in relation to their knowledge and the context of reference, producing or identifying reasons for validity or invalidity and exhibit (or choose an exhibition) in the forms required by the nature of the object being assessed and its context of reference.

An aspect not to be overlooked is that at the origin of educational failures in mathematics there is often a lack of linguistic competence. In particular with regard to mathematics problems, in the Newman and Clements' (Newman, 1977) hierarchy of errors, reading and understanding are in the first place, that is, the wrong answers derive from difficulties in understanding the text even before strategic and / or procedural difficulties. The ability to read and understand a text, but also to produce it, must be cultivated and developed with specific activities. In doing this, we must certainly take into account the shortcomings related to the use of the language itself, even more today with the plurality of ethnic groups and native languages that inhabit our classes. In addition, we need to consider the ability in reading technique and the difficulties encountered by students in the construction of an adequate representation of the situation of the problem, often deriving from the conflict that arises between what Jerome Bruner (1993) identified as narrative and logical thinking. Once again educational need is not the same for everyone and school should not pretend it is and acting accordingly.

Math Skills do not go out of Date

Mathematics plays a central role in modern culture, so students need to perceive mathematics as part of the scientific endeavor, comprehend the nature of mathematical thinking, and become familiar with key mathematical ideas and skills.

According to Paula Schornick (2010) mathematics reveals hidden patterns that help us understand the world around us. Now much more than arithmetic and geometry, mathematics today is a diverse discipline that deals with data, measurements, and observations from science; with inference, deduction, and proof and with mathematical models of natural phenomena, of human behavior, and of social systems.

Bringing Mathematics to New Science Problems

The first step is therefore to identify some great ideas on which to base the mathematics curriculum. "Generative" ideas, according to Ron Ritchhart (1999), as they allow students to investigate important issues, connect to other topics related to other domains of human knowledge as well as to the experience and life of students and then interesting for them. These are issues, often controversial but on which there is a lot of information and resources to discuss and study and develop the topic. These are the socio-scientific issues, which have a conceptual connection with STEM but which raise issues of a social nature and incorporate other disciplines and domains of knowledge (historical, philosophical, political, financial, ethical and moral-religious) and which is simply unthinkable to address with a mono disciplinary perspective. Socio-scientific issues include, for example, global warming, energy resources, food choices, pollution, research on new materials, gene therapy and the whole range of questions related to the huge issue of human migration. Mathematics indeed can enrich the enquiry on these huge topics with its perspective. Mathematical Literacy for Environmental Awareness considers the possibilities of linking mathematical literacy not only to an individual's capacity to solve personal and local problems, but also to global environmental concerns. Let's think about Computer Science. A meaningful school curriculum should deal with social and political aspects of coding. Students should engage with privacy and data protection, understanding cyberattacks, bots and hacking, how algorithms and automation are changing the future of work and that interest groups are behind this. Dealing with socio-scientific subjects, albeit in a manner appropriate to age and school level, means teaching to recognize detectable evidence and making judgments, understanding how to distinguish scientific from pseudoscientific issues, being able to make decisions and to argue, to take actions on issues relevant to everyday life. All this has to do with functional scientific literacy (PISA 2006) but also with education for active citizenship. Without a good socio-scientific education, the student falls prey to moral panic (Cohen, 1972), one of those dark fears artificially created, often functionally to specific interests, and fueled by social networks, the press and TV.

Old and New Models of Mathematics Teacher Professional Development

Teacher training is an essential point at subject level and global level too. Yet to teach mathematics it is certainly necessary to master the subject, but this does not guarantee the effectiveness in the teaching-learning process.

Moreover many young teachers are concerned above all with having a thorough knowledge of all the topics of mathematics and, to be honest, many training courses, especially in the past, were focused on the disciplinary contents too and then on the best way to teach them.

It is almost obvious that a deep knowledge of the discipline is necessary but not enough and is not a guarantee of being a good teacher. Worse, reducing teacher training to content training is a conception of teachers' profession that relegates the student to a passive role of receiver and assigns the role of content delivery to the teacher. This is a teacher-centered way of thinking where the teacher has the responsibility of studying and / or elaborating the best and most effective delivery strategies for prepackaged contents. In this way, he takes responsibility for the foreseeable failure, in the event that the pupil does not learn. The student for his part often becomes intolerant of his passive role and refuses to learn topics that do not involve him, misbehaving as an expression of rejection and revolt. The teacher therefore has to devise strategies to contain students' discontent and rebellion, thus entering a vicious circle that is perpetuated. What is even more serious with this type of teaching, students have no opportunity to develop competences, neither critical thinking nor the ability to approach a new question independently.

Professional development actions must shift the emphasis on the student, by promoting a rethinking of the assignment and class work. Even those who overcome the classic three-step lesson (where the teacher explains, the student repeats the teacher's words, the teacher assigns a mark), often focus class work on completing the task or on the activity itself. Instead, it would be important to train teachers to design an active work for students, focused at developing understanding and the types of thinking that are important for scientists and mathematicians in particular such as looking for patterns, making conjectures, forming generalization, constructing arguments, building explanations and, instead of accepting the easy answer, identifying the complexity of a topic without trivializing it. (Ritchhart, R., Church, M., & Morrison, K., 2011). Consequently, learning arises from the student's autonomous experiences and discoveries so the curriculum is no longer the orderly arrangement of knowledge already given but the set of experiences that will gradually allow students acquire and access knowledge themselves.

In addition, teaching mathematics has nothing to do with the reproduction of a model and not even with the application of a good recipe for all occasions. Instead, it is necessary for teachers to be trained to take on a habitus of self-reflection in their work, an attitude of research while teaching that takes place on a daily basis throughout the professional life, adapting one's actions to circumstances, in the light of the new awareness gradually acquired, with a view to improving through small steps. If the most important thing to teach is "learning to learn" in the perspective of a lifelong learning process, teachers first should acquire the habit of a continuous professional development and this is valid with reference to pre-service, induction and ongoing professional development.

Moreover, that is why in recent years the idea of community of learners has become so important in the field of professional development (Guida & Cinganotto, 2020). Rather than a solitary reflection, the constant discussion with peers about professional practices is profoundly formative for teachers who can learn one from another. In this way, they can build on the experience of their colleagues and benefit from their point of view. Furthermore, they can become aware of the beliefs that implicitly condition their teaching practices and students' learning and can interpret and attribute meaning to the students' mistakes, by giving them a new value.

Research and evidence from the field of mathematics lets us know how effective teachers of mathematics skillfully integrate a range of instructional approaches to meet the diverse learning needs of their students. (Boykin, A. W. & Noguera P., 2011). *Effective teachers know how students learn, know what students need to learn, know what their students already know, encourage risk taking, create purposeful learning experiences, create challenge.* (Coombes)

Challenge can be created through problems that can be defined as '.... *any task or activity for which the students have no prescribed or memorized rules or methods, nor is there a perception by students that there is a specific "correct" solution method*' (Hiebert et al.,1997). Problems can be the starting points and an ongoing means for students to investigate and understand conceptual ideas so they can develop skills and procedures. Through such problems, all students can develop gradually the ability of problem solving and increasingly more complex skills. (Steel, 2009).

THE MLTV4MATH PROJECT

Premises

This paragraph introduces a research project conducted by the author, researcher in mathematics education, in collaboration with eleven mathematics teachers, the whole mathematics department of a technical-vocational school in the district of Naples (Italy). At the core of this project rests the idea that it is important to nurture Thinking and make it visible so that a culture of thinking can be built and a strong learning community established in school and classroom.

Experimentation stemmed from a larger project. In year 2017 a collaboration between the National Institute for Documentation, Innovation and Educational Research (Indire, Italy) and a research group from "Project Zero", a group of researchers within Harvard Graduate School of Education, have been set. Starting from this collaboration, the MLTV Project (Making Learning and Thinking Visible in upper secondary schools) was activated (Cinganotto, Guida, et al., 2018) (Panzavolta et al., 2018). Main objective of the MLTV was to experiment and localize some innovation proposals from PZ, mostly the combination of two theoretical frameworks MLV (Making Learning Visible) and VT (Visible Thinking), experimented and validated in USA for years but tested for the first time in Italy within this project. (Krechevsky, Mardell et al, 2013) (Ritchhart, Church, & Morrison, (2011) MLTV involved from the very beginning three Italian schools with 15 teachers and 3 school principals as co-researchers. The criterion for the schools selection was identified to ensure the maximum variety both in terms of geographical position and in terms of type of school. Another criterion was the strong commitment to innovation they showed through the active role they already played as part of the Educational Avant-Garde Movement, a project for school innovation promoted by Indire in year

2014 together with 22 founding schools that it is nowadays a network of more than 1000 schools. Actually, MLTV project promotes a culture of exploration and the construction of critical thinking and it is based on 3 main tools: Thinking Routines (TR), group learning and documentation. TRs are a sort of organizational structure, didactic protocols composed of a set of questions and a short sequence of phases to guide students' mental processes and make visible their thinking. They encourage students to actively engage on a topic, asking questions, taking stock of previous knowledge, connecting new knowledge to previous ones.

Table 1. Sample TRs from the Visible Thinking Project and their prompts

SEE-THINK-WONDER 1. What do you see? 2. What do you think about that? 3. What does it make you wonder?	CLAIM-SUPPORT-QUESTION 1. Make a claim about the topic 2. Identify support for your claim 3. Ask a question related to your claim
CONNECT-EXTEND-CHALLENGE 1. How are the ideas and information connected to what you already know? 2. What new ideas did you get that extend your thinking in new directions? 3. What is still challenging or confusing for you? What questions or puzzles do you now have?	THINK-PUZZLE-EXPLORE 1. What do you think you know about this topic? 2. What questions or puzzles do you have? 3. What does the topic make you want to explore?
WHAT MAKES YOU SAY THAT? 1. What's going on here? 2. What do you see that makes you say that?	HEADLINES 1. If you were to write a headline for this topic or issue right now that captures the most important aspect to keep in mind, what would that headline be?

Group learning is promoted through five interconnected strategies: to increase students' ability to learn together, to design engaging tasks, to facilitate conversations that deepen learning, to form groups intentionally, to plan an effective synergy between individual, small group and whole class work. In PZ's sense, documentation is the practice of observing, recording, interpreting and sharing, through a variety of media, the learning process and products, in order to deepen learning and make it visible. The research question for the present study deals with the use of the TRs and Protocols, mainly used by teachers in classes with their students, as metacognitive tools for teachers themselves. During a protocol - guided discussion dealing with the analysis of documentation from a class activity, eleven mathematics teachers asked to be present as silent spectators. At the end of the protocol, they asked to take an active role in the project, expressing a training need but also making explicit the potential for mathematical education they saw in the project. So the MLTV4MATH project (Guida, 2018) was activated with the group of mathematics teachers, under the coordination of a researcher working in mathematical education for years. A question arises naturally. Why mathematics teachers were so impressed with the framework so they wanted to participate? Originally the TRs and related documentation were not created neither to achieve disciplinary objectives, nor for the acquisition of specific content, nor for the retention of the latter in the long term. Rather, they were born to develop students' thinking, giving them regular opportunities to practice it and guiding them towards high-order forms of thinking and to make such thinking visible. For this reason, they can give excellent results in any disciplinary field, with particular reference to the peculiar aspects, methods, episteme of the different fields of knowledge, or even with an interdisciplinary approach. However, in the TRs their procedural aspect, the sequence of clear and defined steps sounds familiar to the STEM teachers who naturally relate these characteristics to the structure of the experimental method while their function of developing the capacity for logical argumentation and assertion based on evidence and data actually catches the attention of the teachers of mathematics. They have been built and tested to develop a type of deep thinking. In mathematics, for example, "deep learning" means that students are able to see the basic principles of a problem and are not confused by superficial changes such as transforming an equation in a verbal problem or vice versa. Perhaps for this reason the TR most used by Math teachers are those that develop procedural thinking (see fig. 1)

With reference to the project MLTV4MATH, specific research questions were as follow. Can MLTV (in particular TRs and protocols) improve teachers' professional development and metacognition in mathematics? Which is (if any) the specific utility of the MLTV framework in mathematics? Which TRs, used in class with students and supported by regular documentation analysis, are particularly useful for math education purpose? Which particular types of thinking they develop? What particular difficulties, obstacles and points of attention, specificity, and conditions of inapplicability can math teachers encounter?

Methodology: The Figure of the Teacher - Researcher

The research refers to the collaborative research model which pillars are the attention to Teacher Thinking and the analysis of practices for professional development. (Desgagné, 1997) (Magnoler, 2012) (Shoen, 1987).

It pertains also to a participatory approach as the enquiry actively involves different actors from school and research, namely teachers and principals from the schools identified and a researcher. The research model foresees a process of collaborative knowledge building where all the actors contribute in a peer relationship but with specific roles in all research phases: co-situating the research, co-operating and co-producing the result. This provides for a constant interaction and negotiation between researchers and actors of the school.

The project builds on a new, composite and professionally bivalent teacher figure, the teacher-researcher which is the result of the slow evolution of a motivated and capable teacher. Through different active involvement stages, the teacher faces a research training process. Starting from simple experiments this process leads teachers/researchers to gradually collaborate with researchers in the definition of research hypotheses and data analysis, until reaching autonomy of research and ability to produce articles for publication in journals, present their work at national or international conferences, contribute to knowledge in the field of education and share their findings with the scientific community. Specifically, in the context of the project, this process involved choosing a question, gathering data by documenting learning, analyzing the data together through conversation, making connections to other inquiries and planning new steps to continue exploring the research question. Each individual research question was under the big umbrella of the general question of the project: how do learning and thinking, instructional practices, and classroom culture change when teachers share and discuss documentation of student thinking and learning? And it was relevant to teacher daily practice and open-ended. Data analysis consisted in group discussions on a powerful documentation of learning experiences in order to revisit it and share different perspectives to interpret that experience. This kind of teacher is not afraid to expose him/herself to criticism, he/she is always ready to change and is able to use experiences and materials to support a clearly stated thesis. The aim was to have Italian teachers experiment Project Zero protocols and TRs in order to guide them towards the creation of a real community of practice, helping them transform their idea of learning from an individual action into a collective body of knowledge, involving and learning from the others' thinking skills, feelings, emotions.

Main Phases and Results of the Project

The MLTV4MATH project developed in four phases, the first essentially aimed at the theoretical study and in-depth analysis, on the methodological side, of the framework and related tools. The second, more laboratory-oriented, was aimed at experimenting in the classroom with some TRs to understand them thoroughly and become familiar with them. In this phase, teachers made also their first attempts of collecting documentation. The third consisted in the planning and implementation of didactic experiments, each based on a sequence of TRs with specific purposes and in the production of documentation for sharing and group analysis. Fourth phase was a critical review and refinement of experiments and drafting of reports for publication.

The project was structured in 18 fortnightly meetings that took place at school, organized through discussion protocols. This was chosen to concretely identify the community being born with a physical place and nurture a habit of regular professional collaboration and comparison. In the vision of the project, the school must become the place where teachers have the opportunity to discuss, reflect, rehearse and perfect new practices and create a culture of research and professional collaboration. Knowledge, and the form in which it is used, are products of interactions among individuals – the form in which groups sort their experiences and make sense of their world (Putman, Borko, 2000). A key idea from this point of view is the notion of "dialogue community" as a community that shares forms of thought and communication. In other words, math teachers need opportunities to share what they know, discuss what they want to learn, connect new concepts and strategies to their unique contexts and their specific situations in the classroom. (Darling-Hammond &; McLaughlin, 1995), acquire concepts of mathematics teaching that are now generally accepted, such as the didactic contract (Brosseau, 1986). It is also important that this development initiatives was based on intensive training and well-defined and concrete tasks to be completed within class education, focused on knowledge of mathematics, connected to specific standards for student performance and embedded in a systemic context (Supovitz & Turner, 2000). The ultimate aim of teacher training is the educational success of students.

Some video conferences took place in addition to face-to-face meetings, when meeting in person was impossible. During the 2020 pandemic, this has become inevitably the ordinary way of meeting for the research group and even for experimenting with students.

The first meetings of the research group were dedicated to presenting the project as a whole and to declare its aims, to define the theoretical framework, to constitute the group. The figure of the teacher-research was discussed. Group rules were collaboratively identified and set through a special protocol. According to it, participants are given time to reflect and consider aspects of learning that are important when choosing to work and learn in a community. They have some time to reflect and write on sticky notes their personal ideas, first individually, then in pairs and in a small group, through a process of progressive negotiation and adoption of ideas and needs identified as crucial within a community of learners. The final step is to summarize and group the most common ideas, as gradually agreed upon by the groups, on other sticky-notes to be collected on a poster. As a result the collective set of norms turned to be as follow.

To learn well in this group we need:

- respect for roles, times, ideas
- motivation, curiosity

- share experiences and tools useful for the group's growth
- produce good practices and collaborate for their dissemination
- willingness to share ideas and experiences in an authentic way
- find alternative solutions thanks to the inputs of the other colleagues.

After this preparatory phase, each teacher was invited to formulate their own research question which would then be periodically recalled to reconfirm or modify, refine and clarify.

Teachers' individual questions were:

- **Sabrina:** How can I stimulate students' curiosity and the ability to look around to recognize the mathematical patterns in their own real life experience?
- **Vincenzo:** Students often have prejudices towards mathematics: how to overcome this "natural" aversion to discipline?
- **Roberto**: How can I help kids look critically at the things around them?
- **Lucia**: How can I get students used to observing and so develop in them the ability of abstraction which is fundamental for the study of mathematics?
- **Silvana**: Our students do not listen, do not know or do not want to think. Is there an innovative educational model to involve them in the construction of mathematical concepts and can help me face this challenge?
- **Antonella**: what can I do or practice in my lesson to intrigue and make the lesson interesting?
- **M. Rosaria**: How can this type of activity help students develop critical thinking beyond the immediate?
- **Emilia**: Does exploration and content-sharing help students to think?
- **Mario**: How to support pupils in the development of critical thinking instead of based on unfounded opinion?
- **M Consiglia**: How to make the lesson more stimulating without losing sight of the didactic objectives?

At this point, the work got underway and in subsequent meetings, the matrix of TRs and their classification based on functions were analyzed retrieving it on the PZ website. In anticipation of the TR classroom experimentation, the teachers designed a common format for the activity report.

So the phase 2 started. During school-year 2018/19 each participant teacher tried out several TRs, weaving them into their ongoing classroom practice, making video and/or audio recording, picture and transcription of students speeches and collecting their piece of work on paper. In the final part of the year a group of four teachers experimented together, planning collaboratively and observing each other while teaching. This experimentations involved mainly students of 9th and 10th grade and the most common TR chosen were the ones in Fig. 1, linked with topics referred to the four main categories of mathematical content almost everywhere indicated with numbers, geometry and figures, functions and relationships, data and statistics. After that, the group went deeper discussing specific aspects of their work. Special attention was paid to the criterion for selecting students for group formation and the roles given to each of them, how to observe internal dynamics and how to use these observations, if the format per reporting to the group was effective and useful, if particular criterion was needed in order to select the math topics to apply the routine to, among the ones identified by the math department for the 9th grade that are proportions, percentages, statistics, graphical representations, equations. Phase 2 ended with a focus group were teachers generally converged in affirming that the initial expectations were to receive rather operational indications on how to involve students in mathematics lessons and how to bring them to achieve the teaching objectives. In fact, they realized that the work to be done was rather a research job, that there were no pre-packaged solutions. They greatly appreciated the dimension of teamwork and peer collaboration and were intrigued and interested by this new method. Moving forward, ideas were being clarified and they believed they will get new insights for their professional development. They felt that they were evolving and that they were using routines and protocols more confidently.

I still haven't found the answers but I am confident and I go on with great confidence... I like to work in a group, share ideas and compare myself with others.... Group work is fundamental for me.... I like this way of teaching preparing the routines... I must also understand well if I am doing well...I had somewhat vague expectations....We hope that we are evolving. I started with a question mark we were walking in the dark and looking for the light...I want to understand how they catch the essential elements and how they put them together.... See Think Wonder has spurred their imagination a lot, it really brings out the fantasy...STW is the best, the students question themselves...

Teachers were happy to be observed during their classes because they hoped to get suggestions on how to improve the application of the framework. They reported that students are also informed and proud to be observed. They suggested experimenting in pairs in the same class given that observing everyone's classe would be great but was impossible. They also suggested a calendar for these visits.

we can group in two or three for observations... the students are informed and proud to be observed... the result depends on the class context ... I felt free to make mistakes more than the first few times, even if I didn't respect the times or the sequence of phases.

The most important observations concerned the fact that the students engaged in mathematical discussions, each defending a different position. They had never done it, usually they expected "the right answer" from the teacher and then there were no discussions. In addition, two teachers who collaborated to experiment together in class reported having "seen" the moment when a student understood the topic. And they report it as a particularly emotional moment.

I noticed that in this group there were moments of discussion of different opinions. this in math lessons does not happen, the teacher holds the word and accuracy. I went in to sedate and said I'm glad they are compare their ideas.... They like gratification... we were happy that at the end of the TR a group explained a conceptual map on the straight line. At some point they stopped and one said "well then we could start again backwards". He chose a different solution from the whole group and he enlightened himself, he explained it well to everyone and everything worked... His thought emerged ... this is my hope.

The teachers compared their first experiences with TRs by debating the pros and cons of one or other. In particular, emphasis was placed on those that follow a scientist's way of reasoning and those that can encourage the development of the habit of making statements based on facts, evidence and deductions and not on prejudices and hearsay.

I like Claim-Support-Question. There is the recursive mental scheme that a scientist or a researcher uses...which is " I claim something but not by hearsay, by prejudice but by demonstrating based on proof and evidence"... it is a mental attitude that at this age is not yet developed.... they are still related to the world of fantasy and emotions ... it is an attitude that must be developed in the pupils... today parents replace children and for this reason they don't think... The generation of 40/50-year-old has ruined their children.... as well as Connect-Extend-Challenge open problems to further questions.

Finally, the teachers said they believe that the TR can be permanently included in daily classroom practice without replacing other ways of working, once their validity in improving student learning has been demonstrated. They believe that it is not a problem to have to reduce the number of Math topics covered in favor of slower but more effective methodologies.

everything that is good for kids is good for us too...I don't care that it takes time and I don't finish the topics that I have planned...one year is not enough to understand what we managed to do and get from these guys... TRs are useful if you meet very resistant kids, it's a way to start breaking down the barriers ... you can alternate between the traditional methods...

After Christmas holidays phase 3 started focusing on the analysis of documentation through a new protocol which is called "the fishbowl" referring metaphorically to a fish in a bowl, being watched by observers from outside. The "fish" in our case was a teacher reporting his experience and the relate piece of documentation to be discussed in the group, according to some certain steps and fixed times. The fishbowl protocol, as just said, consists in presenting a piece of documentation by a teacher, related to a teaching activity, letting the colleagues the time to silently observe it, then asking questions or making remarks, while the "fish" being silent and listening. In the following step, after collecting all the questions and comments, the "fish" will reply by providing his/her own point of view related to his/her interpretation of the documentation, in light of the insights stemmed from the colleagues' remarks. Finally the colleagues will briefly express what they have learnt from this discussion, that will enrich their professional experience. The protocol gradually brought a deeper and deeper understanding of students thinking and understanding.

Table 2. The fishbowl protocol

Sample meeting agenda
• Review meeting goals, group norms, agenda, and notes or highlights from the last meeting (2 min) • Participants give brief classroom updates since last meeting (3-5 min) • Presenter provides brief context (3 min) • Group asks presenter clarifying questions (2 min) • Group looks at documentation in silence (3-4 min) • Group responds to the following questions while presenter is silent (10-12 min) • What do you see or hear in the documentation? Point to what makes you say that? • What questions does the documentation raise? (presenter doesn't answer questions) • What are the implications for teaching and learning and next steps for the presenter? • Presenter shares his/her take-aways (2 min) • Group members jot down at least one idea to use in their own classrooms and share ideas with the group (4 min) • Debrief protocol and thank presenter.

Then didactic sequences, lasting longer than a single lesson, were designed, generally consisting of three steps. In each of this steps a TR was woven, corresponding to different types of thinking that were intended to be promoted in the students. For example, the first one was intended to introduce and explore new ideas, the second one to synthesize and organize ideas on the new topic, last one to go deeper into understanding.

In order to design the teaching sequences, a format was agreed among participant. Before working on the definition of the format, teachers discussed how to plan, which aspect had to be took in to account. They ended up with the following. First thinking goals have to be clear as well as disciplinary or content goals. The key question to ask oneself is: *what kind of thinking do I want to develop in my students*? Further it is important to:

- Plan opportunities to interact with the thinking and ideas of others.
- Plan how to give and receive feedback.
- Intentionally build groups.
- Plan to document the process and not just the final product to analyze and understand the learning.
- Make thinking accessible.
- Focus on Math thinking and learning.

In particular, with reference to the types of thinking to be developed, the following have been identified according to the Teaching for Understanding framework (Ritchhart, et al., 2011): Observe something carefully and describe it, Build explanations and interpretations, Reason with evidence, Make connections, Consider different points of view and perspectives, Capture the heart and draw conclusions, Ask questions, Discover the complexity and go beyond the surface of things and the ones most characteristic of problem solving: Identify patterns and make generalizations, Generate possibilities and alternatives, Evaluate evidence, arguments, actions, Identify statements, assumptions, biases, Clarify priorities, conditions and what is known.

Later on, a format for the report of the classroom activities was designed too, which included the insertion of elements of documentation like photos of students' comments written on sticky notes or links to short videos of the pupils engaged in class discussions or link to online repository of content.

The final reports, made filling in the common format, and much more the teacher narratives elaborated starting from the reports, turned out to be a rich and interesting reading that highlighted some significant reflections dealing with well-known problematic aspects of mathematics teaching.

Engagement

The school where the research project took place is not an easy one. The high dispersion rate has only been abated in recent years with the arrival of a new principal convinced that it is up to school to get the students involved. This is why teachers are used to find creative ways to engage students and interest them in the lesson. They consider engagement as a prerequisite as no lesson can be done if the pupils are not involved. Sometimes starting from current topics close to the world of teenagers, they take them by the hand to talk about mathematics almost without realizing it. Sometimes the starting point is a discussion on sensitive subjects like in the case of Anna who has created a learning activity on the body mass index linked to the problem of healthy weight, so important for teenagers and connected to a set of problems and discomforts affecting them, including anorexia and bulimia and indirectly body shaming and even bullying. Almost without thinking about it, the students found themselves writing mathematical formulas, talking about ratios and measuring units to calculate. Anna reported that the TR, Connect-Extend-Challenge, with its well defined and progressive steps, helped her to reach her goal. Sabrina pursued the same goal through an activity based on the TR Claim - Support - Question on the concept of individual water needs given the body weight. Through a real world topic and drawing data in a context close to the students, the school one, she leads them to be involved and therefore to make calculation operations willingly (in mind, in writing and with calculation tools). *This calculation activity*, Sabrina said, *is included as a necessary step for the conceptualization.* The TR with its well-established time for each phase gives to students the necessary time for staying thinking about something in order to build a concept, understand it and connect to previous knowledge.

Sabrina used furthermore the TR See Think Wonder during a lesson named "Fruits just in time" where students were invited to carefully observe an image in which the fruits available with reference to the months of the year were represented. In the third and final phase "what questions does it raise?" questions of some interest arose about the sustainability of the market, the production at km 0, the excessive use of fertilizers, finding interdisciplinary links with the economy for the law of supply and demand, and with science for the influence of production chain on global warming. At this point, the students asked to learn more and searches for further information and almost naturally started manipulating data, trying to interpret graphs, percentage, and statistics. The teacher was good at bringing students practically to ask to do math while recognizing its necessity for understanding real problems of the actual society and real life. *This methodology* - Sabrina said - *encourages peer dialogue on mathematical facts, and this, I repeat, is not usual and above all it is almost never solicited artfully, as can instead happen in a lesson, participated as much as we want, in which the demonstration of a theorem or the resolution of an exercise follow a canvas already written and probably recognized as detached from the learner.*

These words naturally lead us to two more interwined aspects: Classroom dialogues/argument, and Real World context.

Mathematics in Real Contexts

The didactic experiences mentioned above were designed starting from real contexts. The construction of meanings in mathematics, as well as that of all sign systems, passes through the recognition of its pragmatic implications. For this reason, it seems relevant to recognize what are the uses in ordinary contexts of the different systems of signs (and rules) typical of mathematics. This relationship between common thinking and thinking in the forms of mathematics has important didactic implications, because it allows a double movement. On the one hand, it allows to base scientific meanings on ordinary ways of understanding; on the other hand, it allows enriching the concepts developed in ordinary use, enabling their use to be more complex than they would have if they were confined to the context in which they developed.

Making Arguments

Students should be encouraged to discuss and explain their ideas in mathematics lessons (Cockcroft, 1982) because moments of dialogue and discussion are important in the construction of mathematical concepts within the class group. The act of attempting express their thought aloud in words helps students to clarify and organize the thoughts themselves (Pimm, 1987). These moments can constitute, if well conducted, "places" in which to bring out the student's thinking and, starting from them and through comparison and verbal mediation, slowly lead the class to build socially shared and more meaningful knowledge. Some elements may also constitute critical points on which the teachers must work such as: the significance of the questions to be asked to the students, the methods of orchestrating the discussions, the participation of the students, the management of relations, the teaching contract (Brosseau, 1986), and the ability to give and receive feedback. More fully developed pedagogical dialogues have not only been shown to assist students' thinking and learning (Mercer & Littleton, 2007), but are also pivotal for developing students' content knowledge in mathematics through oral language use in discussions (Anderson et al., 2011). It is made evident in curricula across the world, that mathematical processes are an integral and important aspect of learning mathematics. Commonly they include aspects like reasoning, explaining and thinking mathematically (Clarke, Clarke, & Sullivan, 2012).

Students, reiterating the use of TR, Silvana said, acquire the habit of affirming their opinion and / or conjecture but also of accompanying it with the reasons for what they have said, without fear of being judged or immediately evaluated negatively.

In the vision of the project he traditional question, respond, evaluate pattern (Cazden, 1988) has been replaced by a "reflective toss" (Zee & Minstrell, 1997), a questioning sequence meant to facilitate and clarify students' thinking. Teacher's first goal is to try to understand students' comments. If he/she gets the meaning then asks" Can you say something more about that?" otherwise he/she says. "Can you say this in a different way?" Further questions push the students to elaborate and justify their thinking to the teacher but mostly to themselves. "What makes you say that?". Ultimately, it is important to consider the TRs not as separate practices but rather as structures that extend the ongoing practice of questioning and listening.

Multiple Representations

In mathematics education, multiple (iconic, linguistic, mathematical) representations of the same concept are used in order to be able to deal with the different aspects and encourage learning of the concept in its entirety. We refer to classical verbal, graphic, manipulative representations but also to dynamic and interactive digital representations or to computational objects. Roberto last year tried out a lesson connecting proportionality and the straight line in the Cartesian plane through a manipulative activity with some dinosaurs of different sizes, using the TR Connect-Extend-Challenge. This year he replicated the activity with the same topic and the same TR but, due to pandemic, using an online platform where students can formulate hypotheses, verbally and also writing in chat style, and then verify by moving the straight line in the plan to see how the parameters in its equation change of consequently and vice versa. Roberto reported that this year the activity was more successful even because he made it clear to students the TRs they are using and the purpose of each step in terms of thinking activated. This contributed to make them aware of their learning process thereby proving the highly metacognitive value of TRs.

Mario also treated the straight line in the Cartesian plane but by relating it to the graphic resolution of the first degree equations. Again he worked on different representations of a concept (verbal, algebraic, geometrical) through class dialogue. Mario created a unit structured in three lessons, each using a different TR, functional to promote a particular type of thinking. *The focus of the entire process,* Mario says, *is on problem solving, (...) a structured reasoning to solve a complex situation, which cannot be obtained with the automatic application of already known procedures, nor with an intuitive approach so discouraging the habit of proceeding by often uncritical attempts.* The first phase was launched from the request for reflection on this short dialogue: Themistocles says to Cyril: " If you give me a coin, we will have the same number ". Cyril to Themistocles: " If you give me a coin, I will have twice as much as yours ". How many coins do the two friends have in total? The chosen routine "What makes you say that?" helps students describe what they see or know and asks them to build explanations. TR takes the form of an entire class conversation around the topic. The tendency initially expressed by some was to provide numerical answers, rather than actually considering in the abstract whether the problem was solvable or not, and whether it allowed more than one answer. Without giving hints about the answer, Mario invited the groups to ask questions and to reflect on the algorithmic aspect of the dilemma, discouraging the approach by attempts. *Reiterating the use of TRs*, Mario said, *math classes were becoming less teacher - centered and I have spoken less and less.*

Another teacher, Vincenzo, chose to work on the circumference and circle properties and took his cue from a single situation, taken from everyday life, to be analyzed in a crescendo of details and implementations through STW. Vincenzo presented an image of a dog tied up in an Italian garden with hedges, a pond with ducks and a building in the middle. It does not spoke of circumference but promotes a free discussion of the students on the image waiting for them to identify the possible movements of the dog by circumventing the obstacles which, given the chain that has around its neck, can only be composed of arcs of circles of different radius. It is noteworthy Vincenzo's patience in listening to the students' observations without making suggestions, letting them come to the geometric considerations by themselves. In fact, at the beginning, the students made several imaginative observations on where was the house, who was the owner and on the traits of his personality deducible from the way his garden was structured. Only later they recognized geometric shapes and posed the problem of which points in the garden could be reached by the dog, drew the map of the garden, highlighted the arches of circumference, calculated the area of sectors. Students' fantastic stories recall Jerome Bruner's narrative thinking which conflict with the logical/procedural thinking has as a consequence the well known difficulties in the construction of an adequate representation of the situation of problems, In this case Vincenzo made this thinking visible and valued it as a mean to involve students and bring them slowly and naturally to see the hidden geometrical structure of the image. *Perhaps this is why,* as Vincenzo reported, *a student said that math classes no longer seems the same, they look similar to the other subjects' lessons.*

Connecting Previous Knowledge

The difficulties that pupils encounter, in general in the study of mathematics, are related to the symbolic language, to the apparent abstractness and easy discouragement that follows student's failures but also to the common vision of mathematics as a compact corpus with strong internal propaedeutic. Often common people are heard to affirm that they cannot understand mathematics because something got wrong in their first grade so they cannot understand nothing later. TRs make explicit the need of recall, recognize and build on the students' previous knowledge. This is all the more important today that most of our knowledge is acquired in informal contexts, and this happens above all for young people. In particular, TR Connect-Extend-Challenge builds new knowledge by rooting it in the previous one and building it as a development and expansion of the latter.

Teacher Thinking at the End of Experimentation

At the end of the experimentation, the research group made a protocol-guided discussion on the outcomes of this phase. The teachers said that adopting the MLTV4MATH framework was certainly a demanding and challenging operation for the teachers. However, they emerged from this experience aware of having participated in something significant for their professionalization, something that would have left a mark, an imprint and determined a change in their way of being teachers. They accomplished what the sociologist Jack Mezirow (2000) calls "transformative learning". The encounter with a new way of thinking and organizing the lesson constituted the "disorienting dilemma" which, according to Mezirow, gives rise to the process of transformation of the perspectives of meaning (frame of references), thus leading to a change of one's point of view and consequently of acting, in this case, teaching. By joining the project, the teachers decided, consciously, to open up to new stimuli by accepting the challenge, the difficulty and the uncertainty related to getting involved and questioning consolidated practices, with the prospect of changing one's vision of teaching thanks to a personal and profound reflection on the experience lived.

The specificity of the framework in inducing a transformation of teaching practice has been identified by the teachers in the help it gives in recognizing and developing the students' talents and the soft skills that will be necessary for them to move smoothly and independently in the world of labor and at the university. Through the framework and its tools teachers have been able to involve all students in class dialogue, both those who are more fragile and less inclined to present facts and opinions and those who are more equipped with tools and expressive skills.

The application of the frameworks, according to the teachers, also helped to bring out the emotional and social learning in the group and transformed the relationships within the classrooms by favoring the building of empathic relationships among students, between students and teachers, among teachers. MLTV4MATH is considered as a great opportunity to stimulate positivity, discovery, involvement, but also as an opportunity to induce students to reflect more deeply and achieve a deeper understanding.

Teachers appreciated class observations carried out by the researcher and by teachers as critical friends. They are accustomed to working alone and this was a moment of great professional growth. Observation in the classroom had an unexpected secondary effect: the students were proud to be observed, happy to work and to express themselves freely and almost on par with the teacher, even the most shy and those with learning difficulties.

The teachers did not want to stop their experiments because of the Covid pandemic and so they modified some activities by the means of free online tools such as video conferencing platforms, interactive whiteboards, content repositories and mobile app for chat thus showing that this methodology, so strongly based on personal interaction, is practicable even from a distance.

The Final Phase: Teacher Narratives

The last phase dedicated to reflection, re-reading and finer interpretation of what has been experimented and learned has been partially completed. In particular, numerous teachers' narratives were made in textual form, while videos that should have shown exemplary experiences in the classroom or interviews with teachers who shared their experience, were not made due to the measures of social distancing. Some journal articles written with a double perspective both of teachers and researcher are in progress and will be shortly submitted for publication or used as study materials in the framework of a forthcoming national teacher in-service training course.

The narratives characterized by the intentionality of carrying out an experimental action but also by the necessarily personal and subjective point of view of the narrator, have been a way for teachers to share their experience and showed their desire to disseminate practices that could inspire others to test themselves with this methodology. Narration is an instrument that Bruner (1993) considers of great cultural value both in cognitive and emotional terms, that can be a bearer of knowledge and values, narrating an experience with rich details and situating it in time and space so that it assumes meaning for others. For this reason Daniel Taylor (1999) claims that each of us is the product of the stories listened to and lived. Narrative thinking is not something artificial or a literary device; it is instead a proper and natural faculty of human thinking (Smorti, 1994). It is used to understand social events and it is fundamental in the frame of didactic innovation because it can facilitate renewal processes precisely with its function of interpreting experience and its ability to give order and meaning to fragmented facts and to elements of experience that are often random and non-generalizable.

SOLUTIONS AND RECOMMENDATIONS

Meeting in presence at school on a regular basis was a winning choice. Although the teachers were already accustomed and easy in distance learning, the meetings in the teacher room seemed more effective than the ones online via video conference. A second point of attention is that some teachers are truly open to innovation and willing to deeply understand new tools and methodology. The risk to avoid is that someone consider TR to be just a tool like another to structure the lesson. On the contrary, they are at the same time a tool, a pattern of behavior that supports specific thinking moves and a sort of organizational structures to guide students' mental processes and encourage active content processing.

FUTURE RESEARCH DIRECTIONS

This chapter was meant to present briefly some of the outcomes of the MLTV4MATH research project. After briefly describing the frameworks of the project, some of the tools experimented with the mathematics teachers, in particular some of the protocols for guiding discussions and didactic sequences making use of TRs, were depicted.

The aim was to help teachers get aware of the importance of group learning as a collective and reflective learning experience, involving socio-emotional, cognitive and affective dimensions, as described and encouraged in Project Zero frameworks. Actually, joining the project turned out to be a very positive experience for teachers: they found out the importance of group learning as a new adventure that can improve their teaching style as well as their lifestyle.

Another aim was to make them aware of the possibility for mathematics education to contribute to build a better world, educating citizens capable of facing the issues of I4.0 and take an active role in our society.

MLTV4MATH project is still ongoing and the future work consists in a deeper analysis of the collected data and in disseminating tools and first results of the project to a larger number of schools. A deeper investigation will be done on the aspect of the documentation, barely touched in the first year of the project, in terms of both collection and analysis.

CONCLUSION

Mathematics can play an important role in building the world of tomorrow as well as the future of humanity. For this to happen, we will have to unlearn to learn because a paradigm shift is required. This means abandoning the effort to teach procedures and the attempt to carry out the entire curriculum but rather to make students face the big problems of the real world, honoring their complexity without trivializing them. Students should be guided to recognize mathematics as a cultural tool that helps them to see the logical structure within those big themes. We will have to rethink the task assigned to the students, making sure that they collaborate in seeking solutions together, practicing democracy and empowerment, develop high order thinking and finally learn how to learn independently for life. A teacher training is necessary but it should take the form of a participatory research in which theory and practice are combined in a continuous reflective practice in class contexts.

REFERENCES

Anderson, N., Chapin, S., & O'Connor, C. (2011). *Classroom discussions: Seeing math discourse in action*. Scholastic Math Solutions.

Boykin, A. W., & Noguera, P. (2011). *Creating the opportunity to learn: moving form research to practice to close the achievement gap*. ASCD.

Brousseau, G. (1986). La relation didactique: le milieu. In Actes de la IVème Ecole d'Eté de Didactique des Mathématiques. IREM Paris 7.

Bruner, J. (2002). *Making stories: Law, literature and life*. Harvard University Press.

Callahan, R. (1962). *Education and the Cult of Efficiency*. University of Chicago Press.

Cazden, C. B. (1988). *Classroom discourse*. Heinemann.

Cinganotto, L., Guida, M., Laici, C., Mosa, E., & Panzavolta, S. (2018) MLTV, rendere l'apprendimento e il pensiero visibili nella scuola secondaria di secondo grado. *EMEM 2018 Conference Proceedings*.

Clarke, D. M., Clarke, D. J., & Sullivan, P. (2012). Reasoning in the Australian curriculum: Understanding its meaning and using the relevant language. *Australian Primary Mathematics Classroom*, *17*(3), 28–32.

Cockcroft, W. H. (1982). *Mathematics counts. Report of the Committee of Inquiry into the Teaching of Mathematics in Schools*. Her Majesty's Stationery Office.

Cohen, S. (1972). *Folk devils and moral panic*. Routledge.

Coombes, C. *What is Effective Teaching of Mathematics?* Generation Ready. Retrieved online at https://www.generationready.com/what-is-effective-teaching-of-mathematics/

Darling-Hammond, L., & McLaughlin, M. W. (1995). Policies that support professional development in an era of reform. *Phi Delta Kappan, 76*, 597–604.

Desgagné, S. (1997). Le concept de recherche collaborative: l'idée d'un rapprochement entre chercheurs universitaires et praticiens enseignant. *Revue des sciences de l'éducation.*

Dubet, F. (1994). *Sociologie de l'expérience.* Le Seuil.

Gardner, H. (1983). *Frames of Mind: The Theory of Multiple Intelligences.* Basic Books.

Guida, M. (2018). Rendere visibili il pensiero degli studenti ed il loro apprendimento in matematica. In La didattica della matematica, strumento concreto in aula. Atti del XXXII Convegno Nazionale Incontri con la matematica Castel San Pietro (Bo).

Guida, M. (2019). La formazione del docente di matematica. *Essere a Scuola, 6*(201), 14-19.

Guida, M., & Cinganotto, L. (2020). Visible Teacher Thinking and Group Learning. In M. Rehm, O. Mealha, & T. Rebedea (Eds.), *Ludic, Co-design and Tools Supporting Smart Learning Ecosystems and Smart Education.* Springer Nature.

Hannon, V. (2017). *Thrive: Schools Reinvented for the Real Challenges We Face.* Innovation Unit Press.

Hiebert, J., Carpenter, T., Fennema, K., Fuson, D., & Wearne, D. (1997). Making Sense: teaching and learning mathematics with understanding. Portsmouth, NH: Heinemann.

Jablonka, E. (2003). Mathematical Literacy. In Second International Handbook of Mathematics Education. Springer. doi:10.1007/978-94-010-0273-8_4

Krechevsky, M., Mardell, B., Rivard, M., & Wilson, D. (2013). *Visible Learners: Promoting Reggio-Inspired Approaches in All Schools.* John Wiley & Sons.

Magnoler, P. (2012). *Ricerca e formazione. La professionalizzazione degli insegnanti.* PensaMultimedia.

Mercer, N., & Littleton, K. (2007). *Dialogue and the development of children's thinking: a sociocultural approach.* Routledge. doi:10.4324/9780203946657

Mezirow, J. (Ed.). (2000). The Jossey-Bass Higher and Adult Education Series *Learning as Transformation: Critical Perspectives on a Theory in Progress.* Jossey-Boss.

National Research Council. (2001). *Adding It Up: Helping Children Learn Mathematics*. Washington, DC: The National Academies Press. doi:10.17226/9822

Newman, M. A. (1977). An analysis of sixth-grade pupils' errors on written mathematical tasks. In M. A. Clements & J. Foyster (Eds.), Research in Mathematics education in Australia (Vol. 2). Melbourne: Swinburne College Press.

OECD. (2006). *PISA Science Competencies for Tomorrow's World*. OECD iLibrary. https://www.oecd-ilibrary.org/education/pisa_19963777

OECD. (2009). *PISA 2009 Assessment Framework. Key competencies in reading, Mathematics and science*. Retrieved May 9, 2020, from https://www.oecd.org/pisa/pisaproducts/44455820.pdf

Panzavolta, S., Mosa, E., Laici, C., Guida, M., & Cinganotto, L. (2018). MLTV, Rendere L'apprendimento e il Pensiero Visibili Nella Scuola Secondaria di Secondo Grado. *Exploring the Micro, Meso and Macro navigating between dimensions in the digital learning landscape EDEN 2018 Annual Conference Proceedings*.

Pimm, D. (1987). *Speaking mathematically*. Routledge.

Project Zero's Thinking Routine Toolbox. (n.d.). Retrieved May 9, 2020, from https://pz.harvard.edu/thinking-routines

Putman, R., & Borko, H. (2000, January - February). What do new views of knowledge and thinking have to say. *Educational Researcher*, *29*(1), 4–15. doi:10.3102/0013189X029001004

Ritchhart, R. (2015). *Creating cultures of thinking*. Jossey-Bass.

Ritchhart, R., Church, M., & Morrison, K. (2011). *Making Thinking Visible*. Jossey Bass.

Schön, D. A. (1993). *Il professionista riflessivo: per una nuova epistemologia della pratica professionale*. Edizioni Dedalo.

Schornick, P. (2010). Looking at High School Mathematics Education From the Inside Out. *National Association of Secondary School Principals (NASSP) Bulletin.*, *94*(1), 17–39.

Smorti, A. (1994). *Il pensiero narrativo. Costruzione di storie e sviluppo della coscienza sociale*. Giunti.

Steel, C. F. (2009). *The inspired teacher: How to know one, grow one, or to be one*. ASCD.

Supovitz, J. A., & Turner, H. M. (2000). The effects of professional development on science teaching practices and classroom culture. *Journal of Research in Science Teaching*, *37*(9), 963–980. doi:10.1002/1098-2736(200011)37:9<963::AID-TEA6>3.0.CO;2-0

Taylor, D. (1996). *The Healing Power of Stories*. Doubleday.

Toffler, A. (1970). *Future shock*. Bantam Books.

Zan, R. (2017). Il ruolo cruciale del pensiero narrativo nella comprensione dei problemi. *Didattica della matematica. Dalla ricerca alle pratiche d'aula, 2*, 46-57. doi:10.33683/ddm.17.2.3

Zee, E. V., & Minstrell, J. (1977). Using questioning to guide students thinking. *Journal of the Learning Sciences*, *6*(2), 227–269.

Chapter 2
Effective Training of the Teachers in Teaching of Mathematics:
A Reflective Professional Development Model for Teachers-as-Designers

Zerrin Toker
TED University, Turkey

ABSTRACT

This chapter aims to provide an in-service teaching model, which is named as a "reflective teachers-as-designers professional development model" based on the literature, designed to give teachers an understanding of teaching that focuses on developing real-life contexts and developing global competencies. First, the conceptual framework for the learning of teachers and their capacity as designers are discussed. Then, content components and process components of the model are presented by providing the rationale behind each element.

INTRODUCTION

The world is changing. Everything that changes in one field in the world creates a butterfly effect in another field. Education also plays a role in this cycle, which is both affected by the change in other fields and triggers changes in them. The relationship is complicated and it is in precession. Thus, many educational programs have been revised in a way to lead this enormous and growing change in recent years. Based

DOI: 10.4018/978-1-7998-3868-5.ch002

on the fact that the foundation of the skills, knowledge, and values needed by both the business world and a developing society will be laid at school, the developments, which focus on the curriculum and the design of education accordingly, also shape the educational policies in many countries in the world. For example, in the Sustainable Development Goals (SDG) of the United Nations, 17 Sustainable Development Goals (SDG) have been directed towards an understanding of education that can develop not only knowledge but also global competencies. In this context, OECD has recently set a target for the measurement of global competencies for the first time in the PISA 2018 exam.

Changes and trends in education also affect mathematics education. Knowledge of mathematics that is learned in school helps individuals to interpret, understand, and solve daily life problems. Therefore, it is necessary to perceive school mathematics as a tool that acts -in conjunction with other courses- in gaining global competencies, and offers support for making the individual a world citizen as a whole; rather than just a list of learning outcomes that acquire knowledge. Students' ability to use mathematics in real-life situations and being literate in mathematics is one of the most important focal points of learning mathematics (De Lange, 1995; Muller & Burkhardt, 2007). For this point of view, new elements should be added to the practices and mathematics teachers should aim to train mathematics literate students and equip them with math-related competencies. Mathematics teachers need to handle the teaching process like designers and design a teaching process to serve this change. Teachers need to think about which big ideas or global issues can be used as a tool, and shape the teaching process on these big ideas. Big ideas can be defined as a rational, conceptual focus on the achievements in the curriculum (McTighe & Thomas, 2003). In more concrete terms, it is the focal point that will support students to find the answer to the question "What will be good for me to learn?" and guide the teacher to design the teaching process.

As one of the consequences of the shift in paradigm, design thinking has become a frequently heard concept in the field of education in recent years. While studies on how to gain 21st-century skills to students are increasing rapidly, how to design teaching processes for gaining these skills in parallel is an important research focus (Lin, Shadiev, Hwang & Shen, 2020). It is hard to equip students with life and career skills, learning and innovation skills, knowledge, media, and technology skills (Trilling & Fadel, 2009) by continuing usual teaching habits and mindset. As Edelson (2002) mentioned, "if the ultimate goal of educational research is the improvement of the education system, then results that speak directly to the design of activities, materials, and systems will be the most useful result." (p.119). As an adaptation to this perspective, there is a need for teachers to position themselves as the designers of the teaching by bringing together many elements rather than being the exact implementer of a written curriculum.

Design thinking is a methodology to explore the solution of wicked problems (Cross, 2001). Like many other areas, educational practices also involve many wicked problems, and teachers are professionals who are directly exposed to such problems. In this context, "design thinking has the potential to be a useful learning and developmental tool for teacher professionalism" (Nelson & Anetta, 2016, p.12). As Dunne and Martin (2006) mentioned, "Under a design-thinking paradigm, students would be encouraged to think broadly about problems, develop a deep understanding of users, and recognize the value in the contributions of others" (p. 512). Considering that the learners here are teachers, it can be said that design thinking has the potential to be an important auxiliary tool for teachers to solve these problems.

According to some researchers, while design thinking is defined as a pedagogy (Luka, 2014), for others, it is an approach that directs the methodology (Edelson, 2002) The inclusion of design thinking in the teaching process brings along a paradigm shift that enables teachers to create a new perspective on teaching (Cook & Bush, 2018). Thinking teachers-as-designers involves changing the perspective and different understanding of the learning and teaching approaches. Besides, competence is far more than a discipline alone can provide. Therefore, it is important and necessary to address learning and assessments with a holistic perspective, focusing on global competencies, and providing real understanding. In an interdisciplinary manner, students can gain such a holistic understanding of global issues.

Designs expected from teachers regarding instruction in innovative learning environments in which students are actors of learning include;

- Learning the conceptual information specific to a discipline,
- Integration of learning with other disciplines,
- Linking the learned to real life,
- Ensuring consistency between previous learning and new learning,
- Creating teaching processes that take into account individual differences,
- Preparing the ground for collaborative work,
- Supporting cultural and intercultural understanding and values.

Including these, as one of the main purposes of teaching mathematics, improving mathematics literacy of students might be considered as the desired big idea of incorporation of design thinking in mathematics instruction process.

Teachers' adoption of changing teaching programs and innovative teaching approaches targeted in these programs can be achieved by guiding and supporting them. Since the most important factor affecting learning in schools is the teacher (World Bank, 2018) with the development of the teacher, practices will also develop. From this point of view, it is prominent that in-service training programs should be carried out on how to teach teachers with knowledge and skills to train students with global competencies and mathematical literacy.

Background of the Study

Research on the concerns of both preservice teachers (Schussler, Stookberry, & Bercaw, 2010) and beginning teachers (Fuller, 1969) showed that they focused on survival and practical techniques and delivery of instruction in their early teaching experiences, rather than students' learning (Schussler et al., 2010). As they gained experience, by reflecting on their teaching, they start to emphasize curriculum and students" learning of the content (Toker, 2016). The lack of focus on students" learning in initial practices (Schussler et al., 2010, Toker, 2016) underlined in the literature addresses that supporting preservice and novice teachers" reflective practices are essential for effective teaching in their future professional life.

When the case is changing the paradigm, experienced teachers may feel as if they are novice teachers. Teachers' beliefs, practices, and working environment shape and direct their implementation. Teachers should be able to understand and appreciate the changes to implement and not otherwise. (Zanzali, 2003). Understanding the curriculum, expectations, approach, and philosophy is also very important for teachers-as-designers. Teachers analyze the curriculum as one of the most concrete starting points to design the teaching process, identify students' interests and needs, and combine these expectations and analysis results with their competencies and approaches to make a design. In this context, teachers' mindset as being a designer brings with it a different view of all these perspectives and the development needs related to this. Teachers need to be supported in various ways to adapt to the changing paradigm. In doing so, it is very important and inevitable to design in-service training with the support of specialists.

Aim of the Study

It is important that teachers need to adapt to changing processes, and need to ensure the success of this development and change process by reflecting on their practice. Alongside this, it also brings new knowledge acquisition as well. In this context, it is important to make arrangements for in-service training for teachers, in the light of a scientific process and providing mentoring support in transforming their achievements into action. In all these aspects, this chapter aims to provide an in-service teaching model, "reflective teachers-as-designers professional development model" based on the literature, designed to give teachers an understanding of teaching that focuses on developing real-life contexts and developing global competencies. First, the conceptual framework for the learning of teachers and their capacity as designers will be discussed. Then, content components and process components of the model will be presented by providing the rationale behind each element. Although this model has emerged by taking into account the national needs, it contains international elements for the professional training of all mathematics teachers who aim to raise individuals with global competencies. In this context, it has the potential to be a framework for teacher development in different countries.

CONCEPTUAL FRAMEWORK

Teachers are Active Learners

As one of the main elements of teaching, teachers need certain knowledge and perspective to design the lesson to be taught to enable students to make sense of mathematical concepts and learn from their contexts (Saenz, 2009). For example, Wubbels, Korthagen, and Broekman (1997) stated that teachers have deficiencies in linking course content with real-life contexts. Therefore, if there is a desire to change and progress teaching practice and enhance students learning mathematics, there is a need to change the way of thinking about mathematics learning processes of students (Thompson, 1984; Trigwell & Prosser, 1996). With the intention of providing quality instruction, teachers should be trained to use several viewpoints to reflect on teaching situations and students'' learning (Toker, 2016).

Learning can be defined in several ways. In this study, teachers' learning is perceived as a construct defined in connection with the cognitivist and primarily constructivist paradigm. Social constructivism suggests that learning is in a social process in which people interact and communicate with each other and their skills are developed (Vygotsky, 1978).The learner is active in the learning process and learning is perceived as the construction of meaning and understanding within social interaction. When the concern is teachers' learning, Warword (2011) claims that teachers' learning is situated. Theories on situated learning emphasize that learners are active in their learning processes (Lempert-Shepell, 1995; Postholm, 2008), and that learners have to use metacognitive skills in these processes (Flavell, 1979). In the learning process of teachers and their development reflection is a key activity (Postholm, 2008). In this model, in order to support design thinking mind shifts (Goldman & Zielezinski, 2016), reflection is considered as a catalyst.

Teachers Have Design Capacity

The participatory relationship of teachers with curricula (Remillard, 2005) is becoming more and more complex day by day with the changing teacher profile. Student's learning needs, the learning environment, the opportunities, and the other elements related to instruction are more important. At this point, how teachers perceive the curriculum and how they reflect on teaching practices should be examined closely. The way teachers think like a designer in the instructional design process is related to how they approach the curriculum. Teachers' main guide in designing instructional processes (Turkish Education Association, 2009), is the curriculum and its relation with teachers in the design of instruction can be explained by the concept of pedagogical design capacity (Brown & Edelson, 2003). Considering a design process (Brown & Edelson, 2003) where teachers actively shape the learning environment using available resources to achieve their goals, an important indicator of design effectiveness is how high the pedagogical design capacities of teachers are. Teachers sometimes use teaching materials in the same way, sometimes adapt them, and sometimes design teaching materials that they create completely by themselves. Accordingly, Pedagogical Design Capacity includes teachers' working with curriculum materials and their ability to use them to perform certain tasks in the classroom (Brown, 2009). At this point, when it is aimed to develop these capacities of teachers, it is necessary to consider the development of different types of knowledge and skills.

The work of Shulman (1986), which is the inspiration for many research studies dealing with the types of teachers 'knowledge, can be used as a starting point for professionals who aim to develop mathematics teachers' design thinking and pedagogical design capacities. Professional education could be organized "to help teachers learn the range of knowledge and skill they need in focused ways" (Ball, Thames, & Phelps, 2008, p.5). Teachers do not use only one type of knowledge in designing instruction. Instead, they use different knowledge types including subject matter knowledge and pedagogical content knowledge (Davis & Krajcik, 2005). Therefore, for the success of a professional development program for teachers, different knowledge types regarding teaching should be included in the process (Sowder, 2007). Within the scope of this chapter, Shulman's (1987) classification of the types of knowledge as content knowledge, general pedagogical knowledge, curriculum knowledge, knowledge of learners and their characteristics, knowledge of educational contexts knowledge of educational ends, purposes, and values and their philosophical grounds and pedagogical content knowledge would be considered. Beyond that, since teachers need to have customized content knowledge (Ball et.al., 2008), the concept of "specialized content knowledge" is considered as well. Specialized content knowledge is "mathematical knowledge beyond that expected of any well-educated adult but not yet requiring knowledge of students or knowledge of teaching" (Ball et.al., 2008, p.9). Developing all these different types of knowledge can serve the purpose of making teachers more competent in instructional design and development of their pedagogical design capacity. At this point, focusing on the development of different types of knowledge can also shed light on the structural construction of professional development activities for teachers-as-designers. In addition to knowledge types aforementioned above, pedagogical content knowledge for disciplinary practices and pedagogical content knowledge for interdisciplinary practices (Huizinga, Handelzalt, Nieveen, & Voogt, 2015) to develop the designer aspects of teachers are also added.

Design Thinking and Heuristics

At this point, the first factor to be addressed in a model that aims the professional development of teachers will be to increase the knowledge and experience of teachers about design thinking. "Formal training in design thinking becomes a requirement in today's environment" (Pande and Bharathi, 2020, p.2) and with this professional development model, teachers will reflect on their learning processes as a designer, while at the same time, they will have the opportunity to reflect on how students with target audiences learn in teaching processes framed by design thinking. In this way, teachers will take an important step towards developing the mindsets of design thinking (Goldman, Carroll, & Kabayadondo et al., 2012) with an experience other than routine practices.

Each element of a model for the professional development of teachers-as-designers will be presented in association with the "design heuristics" (Davis & Krajcik, 2005). The concept of heuristic represents general rule in designing the professional development model, but should not be perceived in terms of specific guidelines. Several empirical research studies are needed to discuss the validity and working of these design heuristics. At this point, the first design heuristic in the professional development model, in which teachers are aimed to develop teachers-as-designers, is "to establish an empathetic link in designing an instruction focused on the student's learning". By understanding the design thinking and reflecting on the students' learning process and their learning processes, teachers-as-designers are expected to transfer their learning to their teaching.

THE REFLECTIVE-TEACHER AS-DESIGNER PROFESSIONAL DEVELOPMENT MODEL

Metacognitive practices that teachers can return to their learning processes can be used to make teachers recognize the skills and competencies. Thus, the second heuristic is that "to develop teachers in design thinking, the awareness of teachers about their knowledge and learning enhanced by reflection process can be used to make them understand how they shape their instructional designs". The professional development model discussed here aims to provide a perspective for reflective thinking both for the curriculum and for its applications for teachers-as-designers. Besides, how this kind of in-service training has an effective role in designing instructional activities, the effect of these activities on students' learning and the role of the mentorship in development of teachers' design capacities in this process is discussed.

The professional development model discussed here includes the elements that should be in a program for teachers to become designers of instruction. In addition, the model provides information on how the process works. In other words, teachers can use the same process and content elements that they experience as a a learner in this professional development process, in the design of their own mathematics teaching processes. In addition to this, utilizing this kind of the reflective program, teachers can deepen their understanding regarding their learning and about the ways of reflecting this to teaching (Wongsopawiro, 2012). At this point, the most important building blocks that undertake the task of strengthening the bridges between the experiences and accumulation that teachers will transfer from their own learning to teaching processes are working in collaboration, carrying out an ongoing and retrospective analysis in the whole design process, making reflection and being supported by the mentor.

Content Components of the Model

Understanding by Design

The first component of the in-service the training model is the **"Understanding by Design" (UBD) model** (Wiggins & McTighe, 2005). UBD is program design and developmental model that focuses on understanding (McTighe & Wiggins, 2004; Wiggins & McTighe, 2005). It can be used as an important tool to make sense of learning (Wiggins & McTighe, 2005). Curriculum and teaching processes designed using UBD are in teachers' teaching preferences (Brown, 2004), in ensuring the permanence of understanding (Meyer, 2006) and student success (Noble, 2011) has a positive effect.

One of the most important features of UBD's more usual and widely accepted teaching processes is its emphasis on backward planning. According to Wiggins and McTighe (2005), teachers should think like an evaluator when designing a lesson or study unit. This way of thinking requires defining the indicators of whether these goals are achieved immediately after setting the goals of learning. After defining the starting point and the destination, it is necessary to design activities and processes that can be used to reach that point (Wiggins & Mc Tighe, 2005). In retrospective planning, instead of starting with the more common acquisition, book, or existing materials, the process starts with the goals and desired results.

Accordingly, in the UBD model;

1. In the first stage, it is expected that great ideas, skills, and content information will be defined and content information will be divided into meaningful learning parts (Bransford, Brown, & Cocking 1999).
2. In the second stage, it is necessary to define a set of assessment tools and processes that will serve to record and report student learning. Assessment can be used here to support students' understanding of their understanding and status in authentic situations and in-class practice.
3. In the third stage, teachers will plan teaching activities that will guide the process and give students different learning experiences (Wiggins & McTighe, 2005).

Studies in which UBD positively influences students' success are found in the literature (Marzano, Pickering, & Pollock, 2001; Noble, 2011). For this reason, the knowledge and experiences for teachers to adopt UBD as a teaching model and use it in instructional design will be provided to them, supported by model teaching, mentoring process and reflective thinking.

Context Based-Instruction

Another component of this in-service training model is **context-based instruction**. As in the case of several disciplines, for mathematics, context-based education provides rich learning opportunities for teachers-as-designers. Context can be situation that helps students to make sense of facts such as scientific rules. Context-based instruction aims to make sense of the concepts to be learned by linking them with the practices in their lives and supports process skills and conceptual learning (Glynn & Koballa, 2005; Johnson, 2002). It contributes to the success (Hughes, Bailey, & Mechur, 2001) and motivation (Özdemir & Üzel, 2011). However, it is difficult to define contexts for mathematics and to teach mathematics within these contexts (Little & Jones, 2007; Sullivan, Zevenbergen & Mousley, 2003; Vappula & Clausen-May, 2006).

In finding the contexts that can be addressed in designing teaching for mathematics discipline, it can be seen what mathematics issues are related to each other, with the concepts in other disciplines or real life. At this point, it is possible to consider these associations at three levels as (1) horizontal and vertical associations within the discipline, (2) associations with other disciplines and (3) associations with real life. Figure 1 illustrates the inclusiveness of these associations.

Figure 1. Three-level analysis model in determining context for teaching

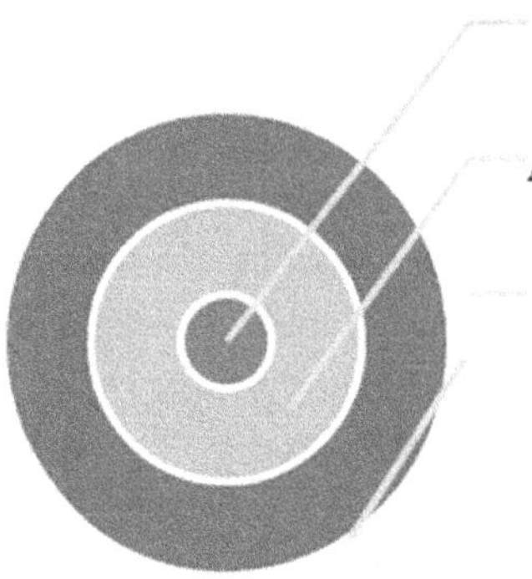

Let's explain this association and contextualization process with a learning process from middle school mathematics:

Although the concept of proportion in teaching process takes place as a concept at middle school level, many related concepts are at previous grade levels. At this point, the first stages of a possible instructional design process for teaching the concept of ratio is determining the prerequisite concepts associated with the concept of ratio. Determining the relationship between this concept and subsequent learning is important since it provides the opportunity to design teaching that can prepare the ground for subsequent learning. Another contribution can be seen as providing support to students' learning needs at different levels of learning. In this context, similar to the analysis process for pre-learning, it can be carried out for future learning. Therefore next step for this is the analysis part. For example, the relationship between graphs related to linear equations and the concept of proportion can benefit both in differentiating and creating a context in teaching the concept of proportion. Within the continuum of the curriculum, the analysis process can continue with analyzing the applications of the rate concept in other disciplines (for example density, the scale of a map) and analysis of real-life analysis examples.

In the professional development model discussed here, teachers-as-designers are expected to analyze the connections for the teaching of the main concepts and contents in the curriculum and to draw the relation maps of the concepts at the three levels to increase the knowledge about the context and to provide support to design mathematics teaching processes using context. It is possible to use Watanabe and Ischinger's (2009) categorization including (1) individual contexts (2) professional contexts (3) social contexts (e.g. sharing) and (4) scientific contexts in defining the context ideas that teachers will obtain by analyzing the associations. Further, teachers would be familiar with different contexts throughout design processes by the help of the model teaching, mentoring process and reflective thinking in the process.

Interdisciplinary Teaching

The third element of the in-service training model is **interdisciplinary teaching**. Interdisciplinary teaching can be defined as to plan the operation of different disciplines for a common purpose, rather than transferring the knowledge of a particular discipline to students. Interdisciplinary practices contribute to students' understanding of learning processes since skill acquisition is a process that needs to be handled with a more complex and interdisciplinary understanding than learning concepts in a single discipline (Aybek, 2001; Yıldırım, 1996). In such an understanding, instead of thinking limiting the concepts and learning of one discipline, the focus would be a big idea.

Research studies in different disciplines reveal an increase in students' academic achievement as a result of teaching based on interdisciplinary interactions (e.g. Gouvea, Sawtelle, Geller & Turpen, 2013). In this context, it is indispensable for designer teachers to associate their teaching focus with teaching focuses on other disciplines. However, when it comes to content of another discipline, teachers cannot be expected to have the same knowledge. Rather, teachers can be expected to recognize what elements they can relate to in their discipline in their learning focus in other disciplines and to evaluate possible integration opportunities. Therefore, teachers must integrate their knowledge (Davis, 2004). In such a collaborative environment, teachers will have the opportunity to benefit from each other's ideas and practices. In addition, they will have a multiplier effect on reflective thinking on their designs. Teacher design teams consisting of the experts of different disciplines will contribute to the development of skills, competencies and other elements of learning of each other, as well as to students' knowledge by serving a higher goal than each discipline can bring.

Differentiated Instruction

Classrooms consist of students whose skills, prior knowledge and interests are different from each other, and teaching should respond to student diversity in this respect (Tomlinson & Kalbfleisch, 1998). As one of the approaches that can best respond to student differences in the classroom, **the differentiation approach** in instruction (Tomlinson, 1999; Tomlinson & McTighe, 2006) is an important element of in-service training model discussed here. Differentiated instruction highlighted that one style of teaching will not mediate every student, especially when there is an incompatibility between the teacher's style and student's style (Levine, 2002).

Besides, differentiated instruction provides teachers the opportunity to vary "learning activities, content demands, modes of assessment, and the classroom environment to meet the needs and support the growth of each child" (Thousand, Villa, & Nevin, 2007, p. 9).

Differentiation in teaching positively contributes to the learning process and student achievement (Lewis & Batts, 2005; Suarez, 2007; Tieso, 2005; Tomlinson, 1999). However, when the applications in the classes where differentiation is applied in teaching are considered, it is seen that four principles stand out (Tieso, 2003; Tomlinson, 1999). These are (1) focusing on the necessary ideas (big ideas) in each learning area, (2) responding to student differences, (3) integrating assessment and teaching, and (4) adapting content, processes, and products to students. In this context, differentiation and UBD in teaching are complementary and supportive approaches (Tomlinson & McTighe, 2006). Tomlinson (1999) defined characteristic ways of differentiating in teaching. According to this; differentiation in teaching involves differentiating the teaching process, activities, assessments, and learning environments, taking into account the students' readiness, interests, learning styles (Hoover & Patton, 2004; George, 2005; Lewis & Batts, 2005; Nordlund, 2003; Powell & Napoliello 2005). Such elements are important to guide the process in the first stage of instructional design. In this context, it is important to provide models on how to increase the knowledge of teachers-as-designers about differentiation in teaching and how to integrate the elements for differentiation in design while designing teaching. For this reason, in the model discussed here, it is seen worthwhile to work on examples of how teachers can increase their knowledge about differentiation in teaching and how they can incorporate differentiating elements into their teaching processes.

Assessment

Another factor that is addressed in the professional development model is the **assessment**. In many countries, teaching methods and assessment and evaluation methods are not parallel (Vos & Kupier, 2005). In this context, it is necessary to equip the teachers-as-designers with elements parallel to the criteria that international assessments also address. Medrich, Calderon and Hoachlander (2002) predict that a teaching process based on real-life contexts do not only cover cognitive goals, accordingly, (1) attitude learning outcomes such as motivation, self-regulation, independent and collaborative learning, (2) learning outcomes for behavior such as the focus on learning and participation, and (3) learning outcomes for skill acquisition such as critical thinking, problem-solving are the outputs that should also be considered in the evaluation of instructional designs. Thus, the use of different assessment processes and tools for knowledge and skills acquisition should contribute to the design. Teachers who set out to acquire 21st-century skills need a different model for the assessment of mathematics learning. Besides cognitive acquisitions, evaluation processes for supporting and evaluating other acquisitions that play a role in the teaching process are included in the professional development model of teachers-as-designers.

Values

The societies want new generations, which are different from the old, equipped with different mathematical knowledge and skills (Bishop, 1999). Self-control, self-confidence, determination, leadership, empathy, communication, original thinking, critical thinking can be given as examples of these competencies and skills (Sam & Ernest, 1997). Thus, in addition to producing information, it is revealed that the affective dimension should not be neglected in education considering that the produced information should be used for the benefit of humans (Ulusoy & Dilmaç, 2016). In this regard, values defined as the development of the knowledge, skills, and abilities (Ryan & Bohlin, 1999) necessary for students to make reasonable choices in their lives are another component of this in-service training model. In addition to this, the basic elements of teaching such as materials development and classroom management, which do not appear to be mentioned here, are not addressed separately as they are the basic and related aspects of the in-service training model and are common parts of the whole elements.

In the light of components discussed above, another design heuristic that takes place in the professional development process of teachers-as designers is "to equip them with several components of effective teaching for new generations".

Process Components

The reflective-teachers-as-designers model that aims to ensure professional development of teachers-as-designers should not only determine what subject teachers will develop but also provide information about how to do this. Therefore, in addition to the content elements mentioned above, some process elements are given below for the model discussed.

Needs Analysis

Analyzing the needs of teachers focusing on developing the role of the designer are included in the model. Zakaria and Iksan (2007) suggested that in-service programs should be directed toward meeting the needs of teachers. Similarly, according to Sowder (2007), professional development should be motivated to provide the knowledge and skills teachers need. Therefore, the identification of perceived needs is crucial for teachers. Teachers may have learning needs for different stages of the design process or for different types of knowledge related to mathematics teaching. Learning process that will meet the needs of teachers in their own learning processes can be provided through a detailed analysis of the needs of the teachers such as design-oriented thinking, design cycle or teaching-related information, and professional development process designs that will be made in line with determining the needs that differ with common needs. Teachers (just like all learners) can ask themselves the question of what it will do to learn. At this point, a design based on the needs of teachers and aiming to meet their needs will bring them closer to finding the answer to this problem.

Besides, in a model where the professional development of teachers is targeted, every study on the process can be seen as a learning experience opportunity on how to design a learning process. Teachers can inspire such a process and adapt it to the level of the class they teach. For example, they can evaluate their students' needs, readiness, and interests as preliminary analysis in instructional design just like their learning process. In this way, teachers also have the opportunity to develop their knowledge of learners and their characteristics (Shulman, 1987). Some questions such as given below can be asked for the design of particular teaching sessions to determine the readiness of teachers about the instructional design and their needs for the types of information required to design.

1. Which grade level(s) include this topic in the mathematics curriculum?
2. Which pre-requisite knowledge is required for the learning of this concept?
3. What basic concepts and procedures does this topic include?
4. Which concepts and procedures in this subject are connected to the other concepts in mathematics? Which concepts are related to the concepts in other disciplines? Which concepts are connected to daily life?
5. Why is it important to learn this topic?
6. Which skills can be supported through learning this topic? How?
7. Which contexts can be used in teaching this topic? How?
8. What types of difficulties do students have while learning this topic?
9. What are the major misconceptions and general mistakes of students regarding this topic?
10. Which methods, representations, strategies, etc. are appropriate for teaching this concept? How and when can these methods, strategies, and representations be used?
11. Which values can be supported in learning this topic? How?

The questions here are given as examples solely for the identification of teachers' needs regarding the design of instruction from the aspect of the content. In addition to these, questions regarding the instructional design process can also be created.

Collaboration

Teachers' designing in collaboration has the potential to be effective in increasing teachers' pedagogical design capacities. Learning has both individual and social aspects (Cobb, 1994) and this is valid for teachers' learning as well (Borko, 2004). At this point, while working in collaboration, teachers will examine the sample materials designed by other teachers, examine the materials designed by others together, evaluate whether these are suitable for teaching purposes, and carry out the instructional design together, making the experience transfer process natural and continuous. These actions will also disrupt the design process (DeKetelaere & Kelchtermans, 1996). On this occasion, the reflections made by teachers together can support this process and provide more qualified products (Nieveen, 2009).

An effective method that can be used to maximize interdisciplinary integration of teachers is to enable them to work collaboratively. At this point, teacher design teams (DeKetelaere & Kelchtermans, 1996; Simmie, 2007) can be used both in teaching practices in schools and in models targeting their professional development. By bringing together teachers from different disciplines to transfer their knowledge and experience related to their discipline to a collaborative learning environment, a natural learning environment will be provided for the development of interdisciplinary field knowledge. In such a professional development model, teachers will have the opportunity to learn by designing, reflecting on the design process together and by transferring their learning experiences.

Mentoring

Mentors have an important role in the professional development of new teachers. The effectiveness of such a process can be mentioned for all professionals who are in a new learning process. In this context, the idea of being a teacher-as-designer is a new learning journey for experienced teachers as well as for the new teachers. Therefore, it can be described as a new learning process since it contains a new perspective and a process of change. Teachers can improve themselves by exchanging ideas with their colleagues in designing, applying, developing, and improving in their classrooms.

Professional development should be intensive, continuous and relevant to implementation (Darling-Hammond, Chung Wei, Andree, Richardson & Orphanos, 2009). That is, teachers need to be supported in the process of their teaching practices

(Ball, 2000), in addition to facilitating form mentorship at the design level. Thus, the support of a mentor should be provided not only in the design phase but also in the execution of all these processes. By this way, teachers can be adapted to the transition to be a teacher-as-designer more smoothly and accurately. Reflective dialogs to be made with the mentor, especially at different stages of the design cycle, can support the effective execution of this process.

Model Teaching

In addition to mentoring, another factor to support teachers is model teaching in the process of becoming a designer. Studies have shown that, model teaching sessions are effective in teachers' professional development (Garet, Porter, Desimone, Birman, & Yoon, 2001). Therefore, they can be used in developing teachers' pedagogical design capacities in this type of professional development model. With the help of model teaching processes to be carried out by experts who have knowledge and experience in instructional design, teachers can find an opportunity to develop a holistic perspective on each stage of the design process, preliminary studies, and complementary elements of the process. Farmer, Gerretson, and Lassak (2003) mentioned that, professional development activities help teachers to learn new ways regarding teaching and learning. Thus, utilizing model teaching activities, teachers-as-designers can learn about teaching and design elements that they will reflect on their practices. Such model teaching can be described as a kind of "templates for instruction" for teachers. Teachers can transfer their methodological knowledge to teaching processes with a different focus.

Reflection

Reflective thinking as an element to serve this purpose is included in the teacher education model discussed. Because reflective thinking is one of the concepts that should be addressed in teacher education first (Tabachnick & Zeichner, 1991) and it is one of the important elements of professional development of teachers (Pollard et al., 2005). Regarding reflective thinking, although a common definition has not been reached in the literature, the most widely accepted definition of Dewey is "Active, persistent, and careful consideration of any belief or supposed form of knowledge in light of the grounds that support it and the further consequences to which it leads" (Dewey, 1933, p.9). However, reflective thinking for teachers is actually "thinking about teaching" (Artzt & Armour-Thomas, 2002, p.7).

When we look at the features that a reflective teacher defined by Zeichner and Liston (1996), it is seen that reflective teachers should have the ability to examine the problems in teaching and try to solve them. These processes can be seen as reflecting the cyclical nature of the instructional approach, where teaching is seen as design, and teacher is seen as a designer. In other words, teachers efforts of designing the teaching, examining the teaching process, determining the problems, improving their designs for the disruptive aspects and making the design better is actually a whole effort to complete the design cycle. The reflective thinking practices of the teacher throughout the entire design process serves as a catalyst to increase the effectiveness of this design cycle or to make the design more qualified.

Therefore, in this model, which is intended to ensure the professional development of teachers-as-designers, teachers constantly reflect on both the design process and their development. Here, when it comes to an interdisciplinary form of instructional design, reflective thinking processes that teachers will cooperate and interact with will provide a multiplier effect. Teachers will have the opportunity to look at different perspectives on their practices when they collaborate and work together on reflective thinking. Besides, it has the potential to sharing and questioning instructional designs and processes (Mau & Harkness, 2020).

SOLUTIONS AND RECOMMENDATIONS

It is inevitable that mathematics teachers could receive the ongoing support that needs concrete action for this change to raise current global citizensi The in-service training model presented here was designed with the aim of finding a solution for this problem of mathematics teachers.

Summary for the Rationale of the Model in 5Ws and 1H

Figure 2 shows the summary for rational for reflective teachers-as-designers professional development model.

Research on the professional development of teachers emphasize that professional development studies should be long-term and ongoing in the process (Sztajn, Wilson, Edgington, & Myers, 2013). Besides, modeling application examples, focusing on students' learning, providing opportunities for reflective thinking and systematic evaluation are accepted as important elements of effective professional development processes (Garet et al., 2001). In this context, it is inevitable to prepare an in-service training model designed for teachers following the design cycle and by applying a design-oriented thinking process. In this type of model, we can consider each learning process that focuses on the development of teachers from different perspectives as micro cycles within the holistic professional development cycle.

Figure 2. Summary for the rationale of the model in 5Ws and 1H

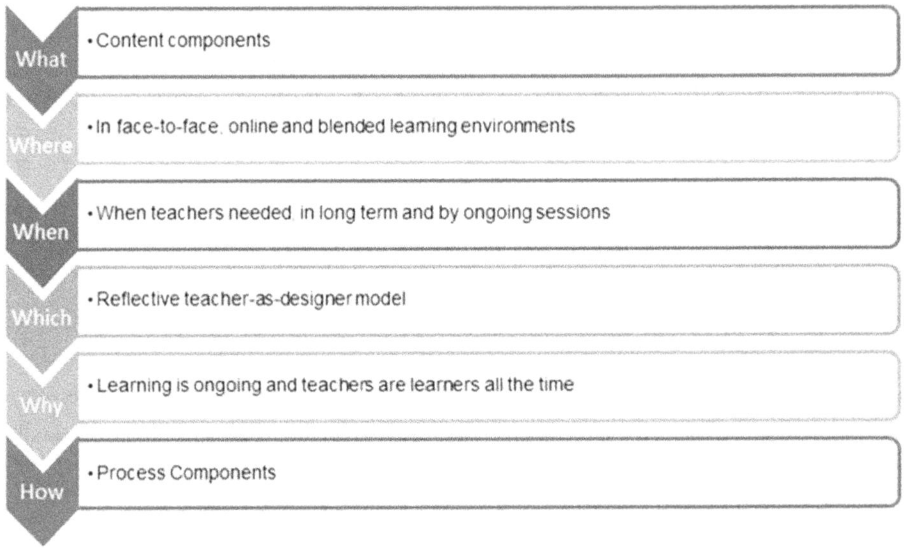

FUTURE RESEARCH DIRECTIONS

It is useful to see the design processes as a research process to see the effectiveness of the designs that teachers perform in their teaching processes in practice. It would help distinguish the parts that work well and do not work, and to guide the following instructional designs in the light of these evaluations. Each teaching process, which takes shape with the reflective thoughts of the teachers, can be considered as a micro cycle for the holistic teaching design that the teacher will realize throughout the year. In this context, it is possible to develop the design thinking and pedagogical design capacity of the teacher by testing the effectiveness of teaching and improving the design throughout the whole process. Thus, this chapter included especially the parts that are related to empathy, ideation and prototyping phases of the design of professional development model and Figure 3 shows the prototype of this model. Therefore, it is open to be upgraded utilizing testing and analysis process.

Figure 3. Prototype of reflective teachers-as-designers professional development model

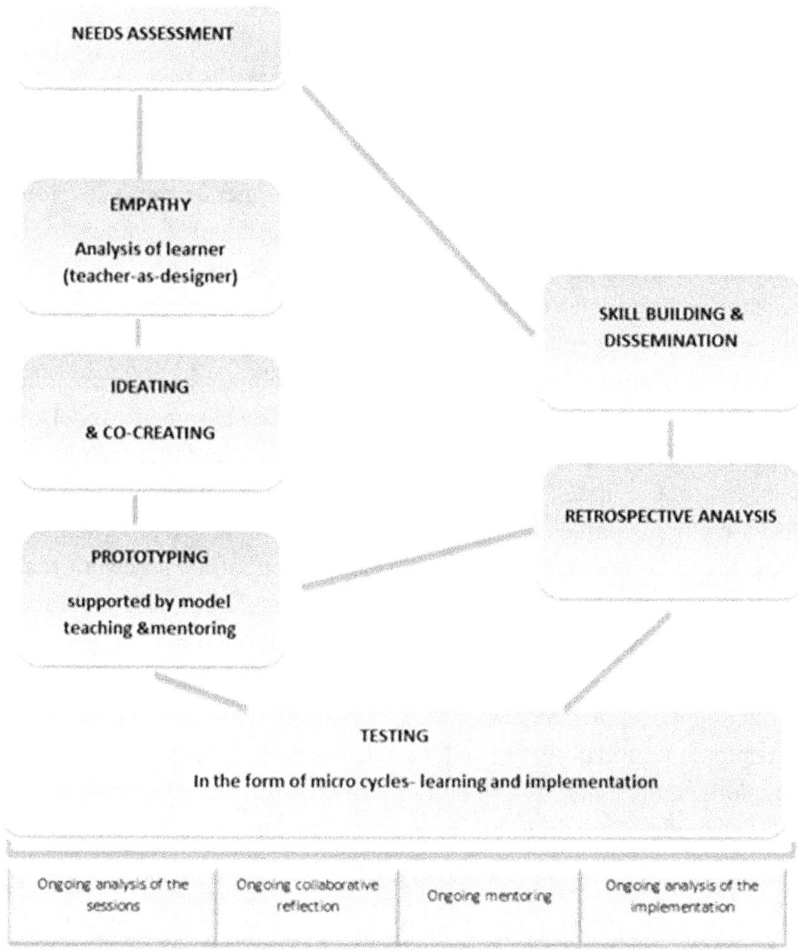

Looking at the instructional design from a reflective perspective also has the potential to offer important support for the teacher to play an active role in his development. As a method to accomplish this Development Experiment (TDE) (Simon, 2000) can be used. TDE for studying the development of teachers was built on the "emergent perspective" (Cobb & Yackel, 1996) which considers learning as a psychological process of the individual and social process of the group at the same time. Emergent perspective, from this aspect, has characteristics of both constructivists teaching experiment methodology (Cobb & Steffe, 1983). The issue in TDE is teacher development. TDE uses the developmental stance of the teaching experiment to study the different aspects of mathematics teacher development at various sites where it occurs.

CONCLUSION

Moving from the mentioned research, as aforementioned above, in this chapter, the professional development model for teachers-as-designers aiming development of teachers' understanding of designing instruction that focuses on developing real-life contexts and developing global competencies is stated and the effects such an in-service training model is discussed through research done in the area of training of teachers in the teaching of mathematics. Design the process is always open to changes and improvement. This design also can be developed through research studies.

The Reflective-Teacher As-Designer Professional Development Model, discussed in this chapter, offers a reflection-supported in-service learning framework in the process of instructional design and professional development for teachers to raise students with global competencies. In this context, content and process components in the model can be considered as the basic elements of a framework that can be used in in-service training of mathematics teachers. The model aims to support the change process of mathematics teachers in order to raise students equipped with global competencies. Besides, to contribute to the development of teachers in different countries, differences in teacher training may occur between countries and the implementation of this model. At this point, differentiation can be included in the implementation processes of the model. In addition, new elements can be added to the model over time due to the practices of different countries and the continuity of change.

REFERENCES

Artzt, A. F., & Armour-Thomas, E. (2002). *Becoming a reflective mathematics teacher*. Lawrence Erlbaum.

Aybek, B. (2001). Disiplinlerarası (bütünleştirilmiş) öğretim yaklaşımı [Interdisciplinary (integrated) teaching approach]. *Eğitim Araştırmaları Dergisi*, *3*, 1–7.

Ball, D. L. (2000). Bridging practices: Intertwining content and pedagogy in teaching and learning to teach. *Journal of Teacher Education*, *51*(3), 241–247. doi:10.1177/002248710005100301 3

Ball, D. L., Thames, M. H., & Phelps, G. (2008). Content knowledge for teaching: What makes it special? *Journal of Teacher Education, 59*(5), 389–407. doi:10.1177/0022487108324554

Bishop, A. J. (1999). Mathematics teaching and values education—An intersection in need of research. *ZDM, 31*(1), 1–4.

Borko, H. (2004). Professional development and teacher learning: Mapping the terrain. *Educational Researcher, 33*(8), 3–15. doi:10.3102/0013189X033008003

Bransford, J. D., Brown, A. L., & Cocking, R. R. (2000). *How people learn: Brain, mind, experience, and school.* National Academy Press.

Brown, J. L. (2004). *Making the most of understanding by design.* ASCD.

Brown, M. (2009). The teacher-tool relationship: Theorizing the design and use of curriculum materials. In J. T. Remillard, B. Herbel-Eisenmann, & G. Lloyd (Eds.), *Mathematics teachers at work: Connecting curriculum materials and classroom instruction* (pp. 17–36). Routledge.

Brown, M., & Edelson, D. (2003). *Teaching as design: Can we better understand the ways in which teachers use materials so we can better design materials to support changes in practice?* Research Report, Center for Learning Technologies in Urban Schools (Northwestern University). Retrieved from http://www.letus.org/papers.htm

Cobb, P. (1994). Where is the mind? Constructivist and sociocultural perspectives on mathematical development. *Educational Researcher, 23*(7), 13–20. doi:10.3102/0013189X023007013

Cobb, P., & Steffe, L. P. (1983). The constructivist researcher as teacher and model builder. *Journal for Research in Mathematics Education, 14*(2), 83–94. doi:10.2307/748576

Cobb, P., & Yackel, E. (1996). Constructivist, emergent, and sociocultural perspectives in the context of developmental research. *Educational Psychologist, 31*(3), 175–190. doi:10.120715326985ep3103&4_3

Cook, K. L., & Bush, S. B. (2018). Design thinking in integrated STEAM learning: Surveying the landscape and exploring exemplars in elementary grades. *School Science and Mathematics, 118*(3–4), 93–103. doi:10.1111sm.12268

Cross, N. (2001). Designerly ways of knowing: Design disipline versus design science. *Design Issues, 17*(3), 49–55. doi:10.1162/074793601750357196

Darling-Hammond, L., Chung Wei, R., Andree, A., Richardson, N., & Orphanos, S. (2009). *Professional learning in learning profession: A status report on teacher development in the United States and Abroad*. Stateford University/ National Staff Development Council.

Davis, E., & Krajcik, S. J. (2005). Designing educative curriculum materials to promote teacher learning. *Educational Researcher*, *34*(3), 3–14. doi:10.3102/0013189X034003003

Davis, E. A. (2004). Knowledge integration in science teaching: Analyzing teachers' knowledge development. *Research in Science Education*, *34*(1), 21–53. doi:10.1023/B:RISE.0000021034.01508.b8

De Lange, J. (1995). Assessment: No change without problems. *Reform in school mathematics and authentic assessment*, 87-172.

DeKetelaere, A., & Kelchtermans, G. (1996). Collaborative curriculum development: An encounter of different professional knowledge systems. *Teachers and Teaching*, *2*(1), 71–85. doi:10.1080/1354060960020106

Dewey, J. (1933). *How we think*. Heath.

Dunne, D., & Roger, M. (2006). Design thinking and how it will change management education: An interview and discussion. *Academy of Management Learning & Education*, *5*(4), 512–523. doi:10.5465/amle.2006.23473212

Edelson, D. C. (2002). Design research: What we learn when we engage in design. *Journal of the Learning Sciences*, *11*(1), 105–121. doi:10.1207/S15327809JLS1101_4

Farmer, J. D., Gerretson, H., & Lassak, M. (2003). What teachers take from professional development: Cases and implications. *Journal of Mathematics Teacher Education*, *6*(4), 331–360. doi:10.1023/A:1026318709074

Flavell, J. H. (1979). Metacognition and cognitive monitoring: A new area of cognitive developmental inquiry. *The American Psychologist*, *34*(10), 906–911. doi:10.1037/0003-066X.34.10.906

Fuller, F. F. (1969). Concerns of teachers: A developmental conceptualization. *American Educational Research Journal*, *6*(2), 207–226. doi:10.3102/00028312006002207

Garet, M. S., Porter, A. C., Desimone, L., Birman, B. F., & Yoon, K. S. (2001). What makes professional development effective? Lessons from a national sample of teachers. *American Educational Research Journal*, *38*(4), 915–945. doi:10.3102/00028312038004915

George, P. S. (2005). A rationale for differentiated instruction in the regular classroom. *Theory into Practice*, *44*(3), 185–193. doi:10.120715430421tip4403_2

Glynn, S., & Koballa, T. R. (2005). *The contextual teaching and learning instructional approach* (R. E. Yager, Ed.). National Science Teachers Association Press.

Goldman, S., Carroll, M. P., & Kabayadondo, Z. (2012). Assessing d.Learning: Capturing the journey of becoming a design thinker. In H. Plattner, C. Meinel, & L. Leifer (Eds.), Design thinking research. Measuring performance in context (pp. 13–33). Berlin: Springer.

Goldman, S., & Zielezinski, M. B. (2016). Teaching with design thinking: Developing new vision and approaches to twenty-first century learning. In L. A. Annetta & J. Minogue (Eds.), *Connecting science and engineering education practices in meaningful ways* (pp. 237–262). Springer. doi:10.1007/978-3-319-16399-4_10

Gouvea, J. S., Sawtelle, V., Geller, B., & Turpen, C. (2013). A Framework for Analyzing Interdisciplinary Tasks: Implications for Student Learning and Curricular Design. *CBE Life Sciences Education*, *12*(2), 197–205. doi:10.1187/cbe.12-08-0135 PMID:23737627

Hoover, J., & Patton, J. (2004). Differentiating Standards-Based Education for students with Diverse Needs. *Remedial and Special Education*, *25*(2), 74–78. doi: 10.1177/07419325040250020101

Hughes, K. L., Bailey, T. R., & Mechur, M. J. (2001). *School-to-Work: Making a difference in education. A Research Report to America*. Institute on Education and the Economy.

Huizinga, T., Handelzalt, A., Nieveen, N., & Voogt, J. (2015). Fostering teachers' design expertise in teacher design teams: Conducive design and support activities. *Curriculum Journal*, *26*(1), 137–163. doi:10.1080/09585176.2014.990395

Johnson, E. (2002). *Contextual teaching and learning*. Corwin Press.

Lempert-Shepell, E. (1995). Teacher self-identification in culture from Vygotsky's developmental perspective. *Anthropology & Education Quarterly*, *26*(4), 425–442. doi:10.1525/aeq.1995.26.4.05x1062v

Levine, M. (2002). *A mind at a time*. Simon & Schuster.

Lewis, S., & Batts, K. (2005). How to implement differentiated instruction? Adjust, adjust, adjust. *Journal of Staff Development*, *26*(4), 26–31.

Lin, L., Shadiev, R., Hwang, W. Y., & Shen, S. (2020). From knowledge and skills to digital works: An application of design thinking in the information technology course. *Thinking Skills and Creativity*, *36*, 36. doi:10.1016/j.tsc.2020.100646

Little, C., & Jones, K. (2007). Contexts for pure mathematics: An analysis of A level mathematics papers. *Proceedings of the British Society for Research into Learning Mathematics*, *27*(1), 48–53.

Luka, I. (2014). Design thinking in pedagogy. *Journal of Education Culture and Society*, *2*, 63–74.

Marzano, R. J., Pickering, D. J., & Pollock, J. E. (2001). *Classroom instruction that works: Research based strategies for increasing student achievement*. Association for Supervision and Curriculum Development.

Mau, S. T., & Harkness, S. S. (2020). The role of teacher educators and university supervisors to help student teachers reflect: From monological reflection toward dialogical conversation. *Reflective Practice*, *21*(2), 1–12. doi:10.1080/14623943.2020.1716710

McTighe, J., & Thomas, R. S. (2003). Backward design for forward action. *Educational Leadership*, *60*(5), 52–55.

McTighe, J., & Wiggins, G. (2004). *Understanding by design: Professional development workbook*. ASCD.

Medrich, E., Calderon, S., & Hoachlander, G. (2002). *Contextual teaching and learning strategies in high schools: Developing a vision for support and evaluation*. Mpr Associates, Incorporated.

Meyer, C. L. (2006). *Learning to teach conceptually: Four preservice teachers' journeys* (Unpublished doctoral dissertation). University at Albany.

Muller, E., & Burkhardt, H. (2007). Applications and modelling for mathematics—Overview. In Modelling and Applications in Mathematics Education (pp. 267-274). Springer.

Nelson, D., & Annetta, L. A. (2016). Creating Disruptive Innovators: Serious Educational Game Design on the Technology and Engineering Spectrum. In Connecting Science and Engineering Education Practices in Meaningful Ways (pp. 3-17). Springer.

Nieveen, N. (2009). Formative evaluation in educational design research. In T. Plomp & N. Nieveen (Eds.), *An Introduction to Educational Design Research* (pp. 89–102). SLO.

Noble, C. L. (2011). *How does understanding by design influence student achievement in eight grade social studies?* (Unpublished doctoral dissertation). Capella University.

Nordlund, M. (2003). *Differentiated instruction: Meeting the educational needs of all students in your classroom.* Scarecrow Education.

Özdemir, E., & Üzel, D. (2011). Gerçekçi matematik eğitiminin öğrenci başarısına etkisi ve öğretime yönelik öğrenci görüşleri [The effect of realistic mathematics education on student achievement and student views on teaching]. *Hacettepe University Journal of Education, 40*, 332–343.

Pande, M., & Bharathi, V. (2020). Theoretical foundations of design thinking – A constructivism learning approach to design thinking. *Thinking Skills and Creativity, 36*, 36. doi:10.1016/j.tsc.2020.100637

Pollard, A., Collins, J., Simco, N., Swaffield, S., Warin, J., & Warwick, P. (2005). *Reflective teaching.* Continuum.

Postholm, M. B. (2008). Teachers developing practice: Reflection as key activity. *Teaching and Teacher Education, 24*(7), 17–28. doi:10.1016/j.tate.2008.02.024

Powell, W., & Napoliello, S. (2005). Using observation to improve instruction. *Educational Leadership, 62*(5), 52–55.

Remillard, J. T. (2005). Examining key concepts in research on teachers' use of mathematics curricula. *Review of Educational Research, 75*(2), 211–246. doi:10.3102/00346543075002211

Ryan, K., & Bohlin, K. E. (1999). *Building character in schools.* Jossey Bass.

Saenz, C. (2009). The role of contextual, conceptual and procedural knowledge in activating mathematical competencies (PISA). *Educational Studies in Mathematics, 71*(2), 123–143. doi:10.100710649-008-9167-8

Sam, L., & Ernest, P. (1997). Values in mathematics education: What is planned and what is espoused? *British Society for Research into Learning Mathematics. Proceedings of the Day Conference held at University of Nottingham*, 37-44.

Schussler, D. L., Stooksberry, L. M., & Bercaw, L. A. (2010). Understanding teacher candidate dispositions: Reflecting to build self-awareness. *Journal of Teacher Education, 61*(4), 350–363. doi:10.1177/0022487110371377

Shulman, L. S. (1986). Those who understand: Knowledge growth in teaching. *Educational Researcher, 15*(2), 4–14. doi:10.3102/0013189X015002004

Shulman, L. S. (1987). Knowledge and teaching: Foundations of the new reform. *Harvard Educational Review*, *57*(1), 1–22. doi:10.17763/haer.57.1.j463w79r56455411

Simmie, G. M. (2007). Teacher design teams (TDTs) building capacity for innovation, learning and curriculum implementation in the continuing professional development of in-career teachers. *Irish Educational Studies*, *26*(2), 163–176. doi:10.1080/03323310701295914

Simon, M. A. (2000). Research on mathematics teacher development: The teacher development experiment. In R. Lesh & E. Kelly (Eds.), *Research design in mathematics and science education* (pp. 335–359). Lawrence Erlbaum.

Sowder, J. T. (2007). The mathematical education and development of teachers. In F. K. Lester Jr., (Ed.), *Second handbook of research on mathematics teaching and learning: A project of the national council of teachers of mathematics* (pp. 157–223). Information Age.

Suarez, D. (2007). Differentiation by challenge: Using a tiered program of instruction in mathematics. In W. Powel & O. Kusuma (Eds.), *Making the difference: differentitation in international schools* (pp. 199–227). ASCD.

Sullivan, P., Zevenbergen, Z., & Mousley, J. (2003). The contexts of mathematics tasks and the context of the classroom: Are we including all students? *Mathematics Education Research Journal*, *15*(2), 107–121. doi:10.1007/BF03217373

Sztajn, P., Wilson, H., Edgington, C., & Myers, M. (2013). Using design experiments to conduct research on mathematics professional development. *Alexandria Revista de Educação em Ciência e Tecnologia*, *6*(1), 9–34.

Tabachnick, B., & Zeichner, K. (Eds.). (1991). Issues and practices in inquiry oriented teacher education. New York: The Falmer Press.

Thompson, A. G. (1984). The relationship of teachers' conceptions of mathematics and mathematics teaching to instructional practice. *Educational Studies in Mathematics*, *15*(2), 105–127. doi:10.1007/BF00305892

Thousand, J. S., Villa, R. A., & Nevin, A. I. (2007). *Differentiating instruction: Collaborative planning and teaching for universally designed learning*. Corwin Press.

Tieso, C. (2005). The effects of grouping practices and curricular adjustments on achievement. *Journal for the Education of the Gifted*, *29*(1), 60–89. doi:10.1177/016235320502900104

Tieso, C. L. (2003). Ability grouping is not just tracking anymore. *Roeper Review*, *26*(1), 29–36. doi:10.1080/02783190309554236

Toker, Z. (2016). *A preservice mathematics teacher's reflective practices on self-improvement regarding teaching and learning process in practice teaching* (Unpublished doctoral dissertation). METU, Turkey.

Tomlinson, C. A. (1999). *The Differentiated Classroom: Responding to the Needs of all Learners*. ASCD.

Tomlinson, C. A., & Kalbfleisch, M. L. (1998). Teach me, teach my brain: A call for differentiated classrooms. *Educational Leadership*, *56*(3), 52–55.

Tomlinson, C. A., & McTighe, J. (2006). *Integrating differentiated instruction and understanding by design*. ASCD.

Trigwell, K., & Prosser, M. (1996). Changing approaches to teaching: A relational perspective. *Studies in Higher Education*, *21*(3), 275–284. doi:10.1080/03075079612331381211

Trilling, B., & Fadel, C. (2009). *21st century skills: Learning for life in our times*. John Wiley & Sons.

Turkish Education Association (TED). (2009). *Öğretmen yeterlikleri* [Teacher Competencies]. Türk Eğitim Derneği.

Ulusoy, K., & Dilmaç, B. (2016). *Değerler eğitimi* [Value education]. Pegem Yayıncılık.

Vappula, H., & Clausen-May, T. (2006). Context in maths test questions: Does it make a difference? *Research in Mathematics Education*, *8*(1), 99–115. doi:10.1080/14794800008520161

Vos, P., & Kuiper, W. (2005). Trends (1995-2000) in the TIMSS mathematics performance assessment in the Netherlands. *Educational Research and Evaluation*, *11*(2), 141–154. doi:10.1080/13803610500110794

Vygotsky, L. S. (1978). *Mind in society: The development of higher psychological processes*. Harvard University Press.

Warford, M. K. (2011). The zone of proximal teacher education. *Teaching and Teacher Education*, *27*(2), 252–258. doi:10.1016/j.tate.2010.08.008

Watanabe, R., & Ischinger, B. (2009). *Learning Mathematics for Life: A Perspective from PISA*. OECD Publishing.

Wiggins, G., & McTighe, J. (2005). Understanding by Design (2nd ed.). Alexandria, VA: ASCD.

Wongsopawiro, D. S. (2012). *Examining science teachers' pedagogical content knowledge in the context of a professional development program* (Unpublished doctoral dissertation). Leiden: ICLON, Leiden University Graduate School of Teaching.

World Bank. (2018). *World development report 2018: Learning to realize education's promise*. International Bank for Reconstruction and Development, The World Bank.

Wubbels, T., Korthagen, F., & Broekman, H. (1997). Preparing teachers for realistic mathematics education. *Educational Studies in Mathematics*, *32*(1), 1–28. doi:10.1023/A:1002900522457

Yıldırım, A. (1996). Disiplinlerarası öğretim kavramı ve programlar açısından doğurduğu sonuçlar [Interdisciplinary teaching concept and its consequences in terms of programs]. *Hacettepe Üniversitesi Eğitim Fakültesi Dergisi*, *12*, 89–94.

Zakaria, E., & Iksan, Z. (2007). Promoting cooperative learning in science and mathematics education: A Malaysian perspective. *Eurasia Journal of Mathematics, Science and Technology Education*, *3*(1), 35–39. doi:10.12973/ejmste/75372

Zanzali, A. (2003). Implementing the intended mathematics curriculum: Teachers' beliefs about the meaning and relevance of problem solving. *Proceedings of the International Conference: The Decidable and the Undecidable in Mathematics Education*.

Zeichner, K., & Liston, D. (1996). *Reflective teaching: an introduction*. Lawrence Erlbaum Associates.

ADDITIONAL READING

Annetta, L. A., & Minogue, J. (2016). *Connecting Science and Engineering Education Practices in Meaningful Ways*. Springer. doi:10.1007/978-3-319-16399-4

Ball, D. L., & Bass, H. (2003). Toward a practice-based theory of mathematical knowledge for teaching. In B. Davis & E. Simmt (Eds.), *Proceedings of the 2002 annual meeting of the Canadian Mathematics Education Study Group* (pp. 3–14). Edmonton, Alberta, Canada: Canadian Mathematics Education Study Group.

Ben-Peretz, M. (1990). *The teacher-curriculum encounter: Freeing teachers from the tyranny of texts*. State University New York Press.

Clandinin, D. J., & Connelly, F. M. (1992). Teacher as curriculum maker. In P. Jackson (Ed.), *Handbook of research in curriculum* (pp. 363–401). MacMillan.

Darling-Hammond, L., & Sykes, G. (1999). *Teaching as the learning profession*. Jossey-Bass.

Design-Based Research Collective. (2003). Design-based research: An emerging paradigm for educational inquiry. *Educational Researcher, 32*(1), 5–8. doi:10.3102/0013189X032001005

Henriksen, D., Gretter, S., & Richardson, C. (2020). Design thinking and the practicing teacher: Addressing problems of practice in teacher education. *Teaching Education, 3*(2), 209–229. https://dschool.stanford.edu/resources and https://tll.gse.harvard.edu/design-thinking. doi:10.1080/10476210.2018.1531841

Koh, J. H. L., Chai, C. S., Wong, B., & Hong, H. Y. (2015). *Design thinking for education: Conceptions and applications in teaching and learning*. Springer.

Lesh, R., & Kelly, A. (2000). Multitiered teaching experiments. In A. E. Kelly & R. A. Lesh (Eds.), *Handbook of research design in mathematics and science education* (pp. 197–230). Erlbaum.

Pande, M., & Bharathi, S. V. (2020). Theoretical foundations of design thinking–A constructivism learning approach to design thinking. *Thinking Skills and Creativity, 36*, 100637. doi:10.1016/j.tsc.2020.100637

Razzouk, R., & Shute, V. (2012). What is design thinking and why is it important? *Review of Educational Research, 82*(3), 330–348. doi:10.3102/0034654312457429

Reimers, F. M., & Chung, C. K. (2016). *Teaching and learning for the twenty-first century: Educational goals, policies, and curricula from six nations*. Harvard Education Press.

Schleicher, A. (2017). *Educating our youth to care about each other and the World*. Retrieved from https://oecdeducationtoday.blogspot.com.tr/2017/12/educating-our-youth-to-care-about-each.html

Sherin, M. G., & Drake, C. (2009). Curriculum strategy framework: Investigating patterns in teachers' use of a reform-based elementary mathematics curriculum. *Journal of Curriculum Studies, 41*(4), 467–500. doi:10.1080/00220270802696115

Wang, S., & Wang, H. (2011). Teaching design thinking through case analysis: Joint analytical process. *Decision Sciences Journal of Innovative Education, 9*(1), 113–118. doi:10.1111/j.1540-4609.2010.00295.x

KEY TERMS AND DEFINITIONS

Design Thinking: Thinking to create solution-oriented strategies for wicked problems.

Mathematics Teacher: Teachers who teach middle or high school mathematics course.

Pedagogical Design Capacity: Competence that includes teachers' use of available resources exactly to plan teaching, adapting and designing a material themselves.

Professional Development: Ongoing process of developing related to the profession.

Teacher Reflection: For teachers, reflection is the process of thinking about their planning, implementation, and evaluation of teaching.

Teacher-as-Designer: Teacher who designs teaching processes for students to learn meaningfully and permanently and to develop a positive attitude towards learning.

Understanding by Design: It is an instructional planning approach that focuses on students' learning by understanding.

Chapter 3
The Incorporation of Big Data in Mathematical Training for the Fourth Industrial Revolution

Jésica Alhelí Cortés Ruiz
https://orcid.org/0000-0002-5459-4874
Instituto Politécnico Nacional, ESCA Santo Tomas, Mexico

Sandra Viridiana Cortés Ruiz
Instituto Politécnico Nacional, ESCA Santo Tomas, Mexico

ABSTRACT

The context of Industry 4.0 is changing the training of mathematical students, new and old generations, in such a way that educational institutions implement strategies and actions to adapt study plans according to the requirements of the new industrial revolution. On the other hand, big data is a cybernetic system that functions as a tool that incorporates mathematical training and implementation and that has recently been included in the educational sphere in order to collaborate with the development of specific competencies based on information technologies and communication, with the purpose of interacting in the intelligent environments proposed by Industry 4.0.

DOI: 10.4018/978-1-7998-3868-5.ch003

INTRODUCTION

The construction of massive data and the dizzying evolution in information and communication technologies invite us to be constantly updated to face change and meet the demands that new generations, globalization and living conditions demand.

According to Mayer and Cukier (2013), a new influenza virus was discovered in 2009. The new strain, which combined elements of the viruses that cause avian and swine flu, was called H1N1 and spread rapidly. In a matter of weeks, public health agencies around the world feared a terrible pandemic would occur. Some commentators warned of an outbreak similar in scale to the 1918 Spanish flu, which affected half a billion people and caused tens of millions of deaths. In addition, no vaccine against the new virus was available. The only hope public health authorities had was to slow down its spread. Now, to do so, they first needed to know where it had already manifested.

In the United States, the Centers for Disease Control and Prevention (CDC) asked doctors to alert them to new cases of the flu. Still, the outlook for the pandemic that emerged was always a week or two late. There were people who could feel sick for days before going to the doctor. Transmitting the information to the central organizations took time, and the CDC only tabulated the figures once a week. With a disease spreading faster and faster, a two-week lag is an eternity. This delay completely obfuscated the public health agencies at the most crucial moments.

A few weeks before the H1N1 virus made headlines, engineers at the internet giant Google happened to publish a remarkable article in the scientific journal Nature. This piece caused a sensation among health officials and computer scientists, but otherwise went largely unnoticed. The authors explained in it how Google could "predict" the spread of winter flu in the United States, not just nationwide, but even by specific regions, and even by states. The company did this by studying what people were looking for on the internet. Since Google receives more than three billion queries daily and archives them all, it had tons of data to work with.

Google took the fifty million most common search terms used by Americans and compared that list with CDC data on the spread of seasonal flu between 2003 and 2008. The intention was to identify those affected by the influenza virus by through what they were looking for on the internet. Others had already tried doing this with internet search terms, but no one had as much data, processing power, and statistical expertise as Google.

Although the Google staff assumed that searches could focus on obtaining information about the flu - by typing in phrases like "remedies for cough and fever" - that was not the point: since they were not sure, they designed a system that did not mattered. All this system did was look for correlations between the frequency of certain information searches and the spread of the flu over time and space. They processed a staggering four hundred and fifty million different mathematical models to test the search terms, comparing their predictions with the flu cases recorded by the CDC in 2007 and 2008. Therefore, they hit a goldmine: Their software found a combination of forty-five search terms that, when used together in a mathematical model, showed a strong correlation between their prediction and the official figures for the disease throughout the country. Like the CDC, they could tell where the flu had spread, but, unlike the CDC, they could do so in near real time, not a week or two later.

So in 2009, when the H1N1 crisis broke out, Google's system proved to be a more useful and timely indicator than government statistics, with their natural reporting lag. In addition, public health officials got an invaluable information tool that is based on big data, "big data": the ability of society to take advantage of information in new ways, to obtain useful insights or valuable goods and services significant.

Derived from the above, in this chapter it is proposed to include Big Data in mathematical education and training as well as in the educational sphere in general, since it is of the utmost importance to link these sectors to be able to develop different competencies demanded by the society of the knowledge in conjunction with the fourth industrial revolution.

In this way, the objective of the chapter is to determine the importance of the incorporation of big data in mathematical training according to the context of the fourth industrial revolution through a documentary research that will be of interest to teachers who are dedicated to mathematics training of students in this area of knowledge.

BACKGROUND

An almost parallel aspect emerged in Germany in 2011, Industry 4.0, which very much coincides with the Big Data theme, since it is characterized by automation, digitalization of processes and the use of technologies in manufacturing. In addition, it considers the personalization of production, the provision of services and the creation of benefit, which implies a combination and exchange of information between people and machines. It is called this way because it is made up of four pillars:

1. The use of data, which is related to the connectivity of systems and the collection of large amounts of information, which requires the use of technologies such as the Internet of things, big data or cloud technology.
2. Artificial intelligence that focuses on the use of systems capable of analyzing information and making decisions for improvement.

3. The human-machine link where the performance of work activities carried out by people is automated incorporating elements of virtual reality or augmented reality, and
4. The transformation from digital to physical which refers to 3D printing for prototypes and final parts after having their behavior simulated by a computer. It also includes the integration of robots to carry out repetitive tasks with little added value.

In this sense, industry 4.0 implies a complete communication network that will exist between companies, social organizations, suppliers, government and clients in order to optimize scarce resources, as well as real time depending on the demands and the state of the associated sections in the network, which will lead to reduced costs and pollution, raw materials and $CO2$ emissions. Thus, Industry 4.0 represents an approach to the innovation of new products and processes, through intelligent processes integrated in work networks throughout the value chain, which gives rise to new forms of collaboration.

In this way, the integration of technologies enables the manufacturing industry to be potentiated with fully integrated, automated and optimized production processes, giving significant results in improving operational efficiency and organizational performance. This situation has generated an impact that is affecting the various areas of the company, from production and organization to research and development, in addition to inventory control, directive management and customer service. Thus, Industry 4.0 has been perceived as a collective term for technologies and concepts of value chain organization where the participants in the value chain offer internal services and interorganizational services.

Thus, Industry 4.0 can be divided into three basic components: 1) Digitization and integration of vertical and horizontal value chains; 2) Digitization of product and service offers; 3) Innovative digital business models.

The first component refers to the “horizontal integration” that occurs between companies through the value chain and the information network to achieve an integration of resources, so that companies can continuously achieve through cooperation and provision of products and real-time services.

For its part, “vertical integration” is based on the use of a network in the manufacturing system, from the demands to the manufacturing services mode, to achieve a personalized production, instead of the traditional fixed production process (for example, a production line). During this vertical integration process, some services support it, such as the integrated digital product. Life cycle management, product design, digital assessment and virtual manufacturing.

The second component refers to the extension of the existing product range with full descriptions of digital products as well as (“embedded systems” / “Internet of things"). These include online connection for regular performance matching and custom product development or wear data.

Finally, the third component consists, in general, of offering internet services based on an oriented service, which is known as reference architecture.

For its part, the principles of Industry 4.0 are as follows:

The first is interoperability, it is exchanging machines and equipment that perform the same function, even from different manufacturers, which gives rise to multiple networks in a trusted environment, for intercommunicating equipment, allowing an awareness that is required for the development of industry 4.0 smart functions.

The second is decentralization. As understood in industry 4.0. It is increasing the capacity of local businesses, personal operations, as well as decision-making machines. Instead of using central computers or passing decisions hierarchically, train and allow local operators to respond to changes and re-adapt, which provides greater flexibility and facilitates the use of specialized knowledge, thus generating a decomposition of the classic hierarchy of production and results in a shift towards decentralized self-organization. The increasing demand for individual products makes more and more centralized and difficult control systems. Thus, embedded computers allow the organization to make the necessary decisions on its own. Only in cases of failure, tasks are delegated to a higher level.

The third is virtualization; it means that using monitoring and machine-to-machine communication, a virtual twin can be abstracted. The sensor data is linked to virtual plant models and simulation models. In this way, a virtual copy of the physical world is created. In case of failure, a human can be notified. In addition, everything you need. Information is provided, such as next work steps or safety provisions. Cyber-physical systems offer to bridge the gap between the exchange of information and the production of meaning by promoting decentralized communication, while two characteristics of Industry 4.0 are fundamental: at the beginning, the activity of the sensors will allow obtaining information about a new level of granularity with the least possible delay. Second, based on real-time data, the simulation will allow anticipating the effect of local optimization in the general context, allowing a better generation of sense and use of decentralized control circuits.

The fourth is the real-time capacity to perform organizational tasks, it is necessary that the data be collected and analyzed in real time, but in Industry 4.0, the notion of real-time capacity goes a step further. It includes plants that can react to the failure of a machine and send products to another machine, as well as a complete link between the final consumers, through social networks or direct sales points, which allows a faster response to changes in the demand. This use of real time.

Information and robotic systems are designed to disrupt modes of production and the way manufacturing is currently organized. This is something that will affect all types of functions, from engineers to maintenance operators, and can also change the physical location of plants.

The fifth is modularity, which involves modular systems that can be flexibly adapted to changing requirements by replacing or expanding individual modules, making adding or removing production modules much easier. Therefore, modular systems can be easily adjusted in the event of seasonal fluctuations or changes in product or production needs, as in the case of the incorporation of new technologies. In addition, many manufacturers' processes, such as product design, production planning, production and engineering and production services, will be simulated as modular and then closely connected end-to-end and interchangeably.

The sixth is service orientation, where business, human and CPS services are available through the Internet of services and can be used by other participants, facilitating the creation of product-service systems. They can be offered internally and across company boundaries. Service orientation and transformation enables organizations to be agile and flexible and respond to market changes much faster than they used to. Enterprise computing is increasingly flexible to partner with the IT networks of its value chain partners to co-create value for consumers (eg, federated framework for service-oriented technology). Organizations can collect and process unprecedented amounts of data with large unstructured data solutions.

In this sense, Ynzunza, Izar, Bocarando, Aguilar and Larios (2017), establish that the implications of Industry 4.0 in business and employment are the following:

In business, it will generate changes in the type of products from physical to digital; higher levels of integration of the organization and the environment; establishment of work networks that create value; infrastructure and shared services in the cloud; intelligent and global supply chains, as well as business models based on cloud services, logistics and distribution.

As for employment, it will allow the creation of new collaboration schemes and social infrastructures; other forms of human-machine interaction; highly specialized job profiles; more complex work process; Assisted and technology-dependent work environments coupled with requirements for the management of digital technologies, robots, programming and data interpretation as transversal skills.

To achieve what is established by the Fourth Industrial Revolution, it is required in the educational training of students to include mathematical literacy, which refers to a set of capacities to develop and apply reasoning in order to solve situations of daily and professional life Therefore, it is oriented towards the acquisition of using the different ways of representing information in the form of models, constructions and graphs to determine the best decision-making in the professional field.

Mathematical competence is the set of capacities to develop and apply reasoning in order to solve situations of daily and professional life, so it is oriented towards the acquisition of using the different ways of representing information in the form of models, constructions and graphs to determine the best decision-making. In this sense, it includes mastering the handling of numbers, measurements and structures to carry out the interpretation of operations and representations of a quantitative nature on personal and professional situations.

Therefore, this competence is relevant because it allows individuals to possess the ability to reason quantitatively to express themselves in mathematical form and to use them as a series of tools for decision-making. Likewise, it includes a positive attitude towards its employment aimed at recognizing its usefulness and validity in the professional field.

Likewise, it has been determined that the development of mathematical and digital literacy are decisive for the formation of skills such as logical thinking and metacognition, which in turn favor the acquisition of the resolution of problematic situations and language in the individual, as well as determining the use of reflection and the ability to discriminate useful information from that which is not (Blanco, 2013; Alterio and Ruíz, 2010).

Consequently, these literacies enable both personal and business success, especially in technologically advanced industries that require the use of numerical knowledge and quantitative principles, as well as the ability to generate, understand, and analyze empirical data to solve complex situations. Furthermore, these capabilities have been confirmed to make technological advancements possible (UKCES, 2011 and Hoyles, 2002).

In the same way, it has been recognized that favoring mathematical skills allows the development of other transversal skills such as commitment, responsibility, decision making, interpersonal communication, flexibility, time management, leadership, creativity and problem solving, work in a team and under pressure (Burns, 2016 and National Soft Skills Association, 2016).

Against this background, HEIs need to reinvent themselves in the context of the Fourth Industrial Revolution because it has been identified that those organizations that are updated with scientific and technological advances are the best positioned because they are characterized by investing in research and development that allows them to generate innovations.

On the other hand, while the concept "big data" itself is relatively new, the origins of big data sets date back to the 1960s and 1970s, when the world of data had just begun with the first data centers of data and the development of relational databases.

Around 2005, people began to realize the amount of data that users generated through Facebook, YouTube, and other online services. That same year, Hadoop, an open source framework created specifically for storing and analyzing large data sets, would be developed. Around this time, NoSQL would also start to gain popularity.

The development of open-source frameworks, such as Hadoop (and, more recently, Spark), would be essential to the growth of big data, as they made big data easier to use and cheaper to store. In the years since, the volume of big data has exploded. Users continue to generate huge amounts of data, but now humans are not the only ones doing it.

With the advent of the Internet of Things (IoT), there are more objects and devices connected to the Internet that generate data on customer usage patterns and product performance. The emergence of autonomous learning has produced even more data. Although big data has come a long way, its usefulness is just beginning. Cloud computing has further expanded the possibilities of big data. The cloud offers truly flexible scalability, where developers can simply streamline ad hoc clusters to test a subset of data. Advantages of big data and data analytics:

Big data allows you to obtain more complete answers, since you have more information.

The availability of more complete answers means greater reliability of the data, which implies a completely different approach to tackling problems.

To understand what "big data" really means, it helps to know some historical background. Here is the Gartner definition, from approximately 2001 (and which continues to be the reference definition): Big data is data that contains greater variety and is presented in increasing volumes and at a higher speed. This is known as "the three words."

In other words, big data is made up of larger and more complex data sets, especially from new data sources. These data sets are so voluminous that conventional data processing software simply cannot manage them. However, these massive volumes of data can be used to address business problems that previously could not have been solved.

The "three words" of big data:

1. Volume

The amount of data matters. With big data, you will have to process large volumes of low-density unstructured data. This can be data of unknown value, such as Twitter data feeds, click streams from a web page or mobile application, or equipment with sensors. For some organizations, this can amount to tens of terabytes of data. For others, even hundreds of petabytes.

2. Speed

Speed is the rate at which data is received and (possibly) used. Generally, the highest data rates are transmitted directly to memory, rather than being written to disk. Some smart Internet-enabled products operate in real time or near real time and require real-time evaluation and action.

3. Variety

Variety refers to the various types of data available. Conventional data types were structured and could be clearly organized in a relational database. With the rise of big data, data is presented in new types of unstructured data. Unstructured and semi-structured data types, such as text, audio, or video, require additional preprocessing in order to derive meaning and enable metadata.

While big data holds great promise, it also faces challenges.

First, big data is characterized by its large size. Although new technologies have been developed for data storage, the volume of data doubles in size every two years or so. Organizations continue to strive to keep up with the growth of their data and to find ways to store it efficiently.

But just storing the data is not enough. To be of any value, the data must be usable, and this depends on its preservation. Having clean data - that is, data relevant to the customer and organized in a way that allows for meaningful analysis - requires a great deal of work. Data scientists spend 50 to 80 percent of their time selecting and preparing data before it can be used.

Lastly, big data technology changes at a rapid pace. A few years ago, Apache Hadoop was the best-known technology used to manage big data. Later, in 2014, Apache Spark would come into play. Today, the optimal approach seems to be a combination of both frameworks. Staying up-to-date on big data technology is a constant challenge.

PROPOSAL

Currently, there is no rigorous definition of big data. At first, the idea was that the volume of information had increased so much that the information being examined no longer fit in the memory that computers use to process it, so engineers needed to modernize the tools to be able to analyze it. That is the origin of new processing technologies, like Google's Map-Reduce and its open source equivalent, Hadoop, which grew out of Yahoo! With them, you can handle much larger amounts of data than before, and that data does not need to be arranged in ordered rows or in the classic tabulations of a database. Therefore, big data will be a source of innovation and new economic value.

Likewise, the educational sphere and the business sector present great ignorance about what massive data means "Big Data"; Today companies do not know what to do with the large volume of data and information stored in different media or databases, which can be of great importance, mainly in decision-making. That is why this research was aimed at demonstrating the importance of Big Data and showing that data is generated with a certain speed and variety, causing growth in volume.

As a Big Data problem, one can contemplate the way data grows today in volume, speed and variety; This is due to the great advance and use of information technologies, and the daily use that people make of them. Forrester, an American market research company that provides advice on the current and potential impact of technology to its clients and the public, defined in 2014, Big Data as the techniques and technologies that make it economical to deal with data on an extreme scale. Big Data is about three things: 1) Techniques and technology, which means that the company has staff, which has great representation and data analysis to have added value with information that has not been managed. 2) Extreme data scale that exceeds current technology due to its volume, speed and variety. 3) The economic value, making the solutions affordable and help business investment". Big Data "refers to the tools, processes and procedures that allow an organization to create, manipulate and manage very large data sets and storage facilities." (Hopkins, 2011).

Salgado (2014) specifies Big Data as "a great volume, speed or variety of information that demands affordable and innovative forms of information processing that allow extended ideas, decision making and process automation". E, he indicates that "Big Data consists of consolidating all the information of an organization and putting it at the service of the business" (Zdnet, 2010).

As an example, a survey conducted by LogLogic finds that 49% of organizations are somewhat or very concerned about managing Big Data, but 38% do not understand what it is, and 27% say they have partial understanding; Furthermore, the survey found that 59% of organizations lack the necessary tools to manage data from their IT systems. Khera explained that: "Big Data is about many terabytes of unstructured data", where "Information is power, and Big Data, if managed correctly, can provide an amount of knowledge ...". According to the survey, 62% mentioned that they had already managed more than one terabyte of data. This indicates that the volume of data in the world is increasing at an almost incomprehensible rate M. Salgado, Oracle is committed to Big Data with technology and projects (Montoro, 2012).

Gartner's Beyer and other experts point out that: "Big data, or Big Data, requires major changes to the server, storage infrastructure and information management architecture in most companies." McKinsey says that it is necessary to prepare to hire or recycle personnel, since companies or organizations lack people trained in Big Data. In addition, "it projects that by 2018, in the United States alone, between 140,000 and 190,000 new experts in statistical methods and data analysis technologies will be needed, including the widely publicized role of data scientist." Catalina Williams points out that "The people who built the databases of the past are not necessarily the people who are going to build the databases of the future."

According to IDC's Worldwide Big Data Technology and Services 2013-2017 study, "Big Data technology and services will grow at a compound annual growth rate of 27% to reach $ 32.4 billion in 2017, about six times the rate growth of the general market for information and communication technologies". (computerworld.es, 2013).

There are three characteristics or dimensions: Volume, Speed, and Variety.

1) Volume: every day, companies register a significant increase in their data (terabytes, petabytes and exabytes), created by people and machines. In 2000, 800,000 petabytes (PB) of stored data were generated and this figure is expected to reach 35 zettabytes (ZB) in 2020. Social networks also generate data, as is the case with Twitter, which alone generates more than seven terabytes (TB) daily, and from Facebook, 10 TB of data every day. Some companies generate terabytes of data every hour of every day of the year, that is, companies are inundated with data (IBM, 2012).
2) Variety: It can be mentioned that it goes very hand in hand with volume, because according to it and with the development of technology, there are many ways of representing data; This is the case of structured and unstructured data; The latter are those that are generated from web pages, search files, social networks, forums, emails or sensor products in different activities of people; An example is converting 350 billion meter readings per year to predict energy use.
3) Speed: It refers to the speed with which the data is created, which is the extent to which software development products increase (web pages, search files, social networks, forums, emails, among others).

The three characteristics are consistent with each other; for example, analyzing 500 million call records per day in real time to predict customer loss (IBM, 2012).

Big Data grows daily, as already mentioned, and one of the justifications is that the data comes from a wide variety of sources, such as the Web, databases, click traces, social networks, Call Center, geospatial data, data semi-structured (XML, RSS), from audio and video, data generated by thermometers, data from website navigation for a certain time, RFID (Radio Frequency Identification - radio frequency identification).

And, there are some benefits of Big Data analysis for organizations, as observed in the marketing area, demonstrated in the survey carried out by TDWI (The Data Warehousing Institute), when it asked: "Which of the following benefits would occur? If the organization implements some form of Big Data analysis? "61% responded that it influences socially; 45%, that there will be more business points of view; 37% favored automated decisions in real-time processes; 29% mentioned that planning and forecasting would be improved, and 27%, that consumer behavior would be understood (Redmond, & Wilson, 2012).

According to the same survey, certain drawbacks of Big Data analysis were found, among them: lack of staff and human resource skills (46%), difficulty in the architecture of a Big Data analysis system (33%), problems with usable Big Data for end users (22%), lack of corporate sponsorship (38%) and lack of a convincing business case (28%), lack of database analysis (32%), Big Data scalability problems (23%), query speed (22%) and difficulty loading data fast enough (21%), among others.

Thus, when asked about how often they perform Big Data analysis, it was found that annually, 15%; monthly, 35%; weekly, 14%; daily, 24%; every few hours, 5%; every hour, 4%; in real time, 4%. The above was the result of 96 interviewees (Redmond & Wilson, 2012).

The goal of data analysis is to examine large amounts of data with a variety of classes, in order to discover information that is relevant and useful for the company, in a way that allows it to make the best decisions and obtain competitive advantages compared to others. of its class, and data analysis is performed with database technologies such as NoSQL, Hadoop and MapReduce, which support Big Data processing.

Big Data Technologies

For data management it is necessary to have two basic components, both the hardware and the software; regarding the first, there are technologies such as Massive Parallel Processing (MPP) architectures, which help quickly to process it. To handle unstructured or semi-structured data it is necessary to resort to other technologies; it is here where new techniques and technologies appear, such as MapReduce or Hadoop, designed for the management of structured, unstructured or semi-structured information.

A. Apache Hadoop

Olavsrud (2012) indicates that Apache Hadoop is an open source software framework for distributed data intensive applications originally created by Doug Cutting to support his work at Nutch, an open source web search engine. Hadoop is now one of the most popular technologies for storing the structured, semi-structured, and unstructured data that make up Big Data. Hadoop is available under the Apache 2.0"license. Likewise, hadoop.apache.org mentions, "it is a software library that allows the distributed processing of large data sets through groups of computers that use simple programming models. It is designed to go from individual servers to thousands of machines, each offering local computing and storage."

Chuck (2011) describes Hadoop as an open source framework, which allows writing and running distributed applications that process large amounts of data. It has some important features:

- It was designed to run on relatively large groups of hardware, that is, on robust clusters.
- It is robust, because in the event of a hardware malfunction it can overcome such situations without major inconvenience.
- It has the advantage of being scalable, which indicates that it allows growing or adding nodes to the cluster with relative ease; For example, given the rapid growth of social networks, it allows adding more nodes with ease.
- It is simple, allowing users to write code efficiently, for distributed software.

Hadoop got its start as a subproject of Nutch, which was itself a subproject of Apache Lucene; it is a text indexing and bibliographic search, that is, it allows searching within documents. Nutch is a more ambitious project than Apache Lucene, the aim is to design a search engine for the web, which contains a parser for HTML, a web crawler, a link-graph database and other necessary additional components.

Currently, Hadoop shows significant advantages over SQL (Structured Query Language) databases, which are presented as a design for the management of structured information, where the data resides in relational tables that have a defined structure, as it was designed for information unstructured or semi-structured, such as text documents, images, and XML files.

"Hadoop can handle all types of data from disparate systems: structured, unstructured, log files, images, audio files, email files, communications, just about anything you can think of, regardless of its native format" (Cloudera, Inc., 2013; Chuck, 2011). It is a framework used to write and run distributed applications that allows to process large amounts of data. Hadoop is made up of two modules: Hadoop Distributed File System (HDFS) and HadoopMapReduce.

1) Hadoop Distributed File System (HDFS: Hadoop Distributed File System): It is a highly fault-tolerant, scalable file system with a distributed architecture; it can store 100 TB in a single file, which is not so easy on other types of file systems. In addition, it gives the appearance of being working on a single file, but what you really have is that they are distributed over several machines for processing.

Chuck (2011) also points out that HDFS was designed for batch processing, rather than interactive use by users; but it was actually designed to solve two important problems that arise in large-scale data processing: the first is the ability to decompose files into several parts and process each one independently of the others, and ultimately consolidate all divisions of the file into one; The second problem was fault tolerance, both at the level of file processing and in a general way of the software, when processing distributed data; what is sought is that the system can recover from the failure that may occur without affecting the process too much (Cloudera, Inc., 2013; Chuck, 2011).

When HDFS was created, three goals were set:

- Allow processing of files with sizes from gigabytes (GB) to petabytes (PB).
- Being able to read data at high speeds.
- Ability to be executed on a machine, without requesting special hardware.

The HDFS architecture is made up of a main node (NameNode) and several slave nodes (DataNodes).

The main node is the master server, dedicated to managing the file name space and controlling the access of the different user files; in addition, the master node is in charge of managing the operations of opening, closing, moving, naming and renaming files and directories.

Slave nodes (DataNodes), as their name implies, represent the slaves of the HDFS architecture. On an HDFS there can be thousands of slave nodes and tens of thousands of HDFS clients per cluster; This is because each slave node can run multiple application tasks simultaneously. The function of the slave node is to manage both the reading and the writing of the users' files, and to perform the replication according to what the master node (NameNode) indicates, (Cloudera, Inc., 2013).

2) Hadoop MapReduce: It is a yarn-based system for parallel processing of large data sets. Hadoop MapReduce is a software framework created in order to make applications that can process large amounts of data in parallel, on the same hardware. When the data enters to be processed, it is divided independently, for its processing, that is, in a distributed manner on different hardware that exists. MapReduce is made up of a master, called JobTracker, and a slave, TaskTracker, for each node. The former is in charge of scheduling the tasks, the components that manage the slave, and the latter executes the tasks as directed by the master.

MapReduce is used in solutions where it can be processed in parallel and, in addition, with large amounts of information, that is, with volumes of petabytes, otherwise it would not be a suitable solution. Another condition is that MapReduce can be used in processes that can be broken up into map () and reduce () operations, which are defined based on structured data. MapReduce was designed as a programming model, so that large data processing could be performed, and, in the same way, to solve the existing problem of scalability. MapReduce "is a programming model for data processing." It can be run in various programming languages, such as Java, Ruby, Python, and C ++ (Zikopoulos, Eaton, DeRoos, Deutsch, & Lapis, 2012).

SOLUTIONS AND RECOMMENDATIONS

As mentioned in previous sections, the fundamentals of Industry 4.0 and Big Data, as well as the relationship between the two, are emerging technologies that aim to develop intelligent systems. In this sense, based on the findings of this chapter, it is suggested that the training of mathematics students and the educational sphere should be focused on information and communication technologies that allow the storage of large volumes of information that are later systematized to fulfill with the optimization of processes and simplification of all the information gathered.

FUTURE RESEARCH DIRECTIONS

Future research directions on this topic may focus on how initiatives and strategies are being developed and implemented in the educational training of mathematics for the development of competencies and how students have benefited from their integration into Industry 4.0.

CONCLUSION

Thus, in the contextual framework of the Fourth Industrial Revolution, a world of continuous changes is envisioned, where science and technology are at the base of the economic competitiveness of countries, where innovation plays a strategic role in competition Therefore, they are forced to meet a series of requirements to compete successfully in an increasingly globalized economy, including high investments in both education and research. With regard to organizations and individuals, the acquisition of new qualities is required to adapt to accelerated changes in the context, which implies continuous or permanent training. As long as countries, organizations or individuals fail to meet these requirements, they run the risk of being excluded from the knowledge society and, with it, from globalized economic competition.

Consequently, Industry 4.0 is generating various transformations oriented towards the digitization of information and production systems for management activities; automation systems for data acquisition of machinery and production lines; with the exchange of information for the monitoring and control of processes and decision-making in real time, which implies that the members of the organization also adapt to these fundamental changes.

In this way, a new perspective of training for employment has been generated, focusing on the management of professional competencies in relation to their development, link with the labor sector and decision-making, so human talent is the fundamental element that It allows generating innovative strategies in companies that allow responding to changing conditions in the markets, as well as fostering not only economic but integral growth in organizations (UNESCO, 2010).

Under this panorama, the Ministry of Economy established a set of strategies in its roadmap for Industry 4.0, called Crafting the future: A roadmap for Industry 4.0 in Mexico, which aims to help the Mexican industry, the different chains supply of strategic sectors, academia, research groups and the government so that they prioritize the trends and necessary technologies that support Industry 4.0.

These strategies are focused on education, technology, digital economy and projects (Ministry of the Economy of Mexico, 2016). Finally, the purpose of this research is to provide a general overview of the generic and specific importance of including big data according to the context of the fourth industrial revolution in order to assist the educational sphere in the design and preparation of content study of mathematics that converge with the requirements demanded by the strategic sectors of Industry 4.0.

REFERENCES

Adelman, S., Moss, T., & Majid, A. (2005). *Data Strategy*. Prentice Hall.

Alterio, G., & Ruíz, C. (2010). Mediación metacognitiva, estrategias de enseñanza y procesos de pensamiento del docente de Medicina. *Educación Médica Superior*, *24*(1), 25–32.

Blanco, R. (2013). *El pensamiento lógico desde la perspectiva de las neurociencias cognitivas* (Tesis doctoral). Universidad de Oviedo, España.

Burns, M. (2016). *Skills for work, skills for life. I can talk 8*. Inglaterra: I CAN. Disponible en https://www.ican.org.uk/about-us/

Camargo-Vega, J., Camargo-Ortega, J., & Joyanes, L. (2015). Conociendo big data. *Facultad de Ingeniería*, *24*(38), 63–77. doi:10.19053/01211129.3159

Carrasco, F. (2013). *Los 6 pasos que su organización debe seguir para confiar en Big Data*. América Latina. Disponible en: http://www. cioal.com/2013/07/31/los-6-pasos-que-su- organizacion-debe-seguir-para-confiar-en-big- data/

Chuck, L. (2011). *Hadoop in Action, Publisher*. Manning Publications Co.

Cloudera, Inc. (2013). Disponible en: http://www.cloudera.com/content/cloudera/en/ why-cloudera/hadoop-and-big-data.html

computerworld.es. (2013). Disponible en: http://www. computerworld.es/sociedad-de-la-informacion/ el-mercado-del-big-data-crecera-hasta-los- 32400-millones-de-dolares-en-2017, 2013.

Dijcks, J. (2013). *Oracle: Big Data for the Enterprise*. Disponible en: http:// www.oracle. com/technetwork/database/bigdata-appliance/ overview/wp-bigdatawithoracle-1453236. pdf?ssSourceSiteId=ocomes

Hadoop.apache.org. (2013). Disponible en: http:// hadoop.apache.org/

Hopkins, B. (2011). *Beyond the Hype of Big Data*. Disponible en: http://www.cio.com/ article/692724/Beyond_the_Hype_of_Big_Data.

Hoyles, C. (2002). Mathematical skills in the workplace: Final report to the Science Technology and Mathematics Council. Inglaterra: Institute of Education, University of London.

IBM. (2012). *Big Data and analytics platform*. Disponible en: http://www-01.ibm.com/software/data/bigdata

Joyanes, L. (2013). *Big Data: Análisis de grandes volúmenes de datos en organizaciones*. Editorial Alfaomega.

Mayer-Schönberger, V., & Cukier, K. (2013). *Big data: la revolución de los datos masivos*. Turner.

Montoro, S. (2012). *Server and Cloud Platform*. Disponible en: http://lapastillaroja.net/2012/02/ nosql-for-non-programmers/

National Soft Skills Association. (2016). *Skills employers seek*. Disponible en https:// www.nationalsoftskills.org/

Olavsrud, T. (2012). *Big Data Causes Concern and Big Confusion*. Disponible en:http:// www.cio.com/article/700804/Big_Data_Causes_Concern_and_Big_Confusion?page=2&taxonomyId=3002

Plugge, E., Membrey, P., & Hawkins, T. (2010). *The Definitive Guide to MongoDB: The NoSQL Database for Cloud and Desktop Computing*. Published Apress Media LLC.

Redmond, E., & Wilson, J. (2012). *Seven Databases in Seven Weeks*. O'Reilly Media, Inc., Pragmatic Programmers, LLC.

Russom, P. (2012). *Big Data Analytics, TDWI*. The Data Warehousing Institute.

Salgado, M. (2014). *Oracle apuesta por Big Data con tecnología y proyectos*. Disponible en: https://www.computerworld.es/big-data/oracle-apuesta-por-big-data-con-tecnologia-y-proyectos

UKCES. (2011). *The supply of and demand for high-level STEM skills*. Briefing Paper. Commission for Employment and Skills.

UNESCO. (2011). *ICT Competency Framework for Teachers*. París: UNESCO. Disponible en http://unesdoc.unesco.org/images/ 0021/002134/213475e.pdf

Zdnet.com. (2010). *Big Data*. Disponible en: http:// www.zdnet.com/ search?q=big+data

Zikopoulos, P., Eaton, C., DeRoos, D., Deutsch, T., & Lapis, G. (2012). *Understanding Big Data*. McGraw-Hill Books.

Zikopoulos, P., deRoos, D., Parasuraman, K., Deutsch, T., Corrigan, D., & Giles, J. (2013). *Harness the Power of Big Data*. McGraw-Hill Companies.

ADDITIONAL READING

Dalenogare, L., Benitez, G., Ayala, N., & Frank, G. (2018). The expected contribution of Industry 4.0 technologies for industrial performance. *International Journal of Production Economics*, *204*, 383–394. doi:10.1016/j.ijpe.2018.08.019

Foro Económico Mundial. (2019). *Fourth Industrial Revolution*. https://intelligence.weforum.org/topics/a1Gb0000001RIhBEAW?tab=publications

Frank, A., Dalenogare, L., & Ayala, N. (2019). Industry 4.0 technologies: Implementation patterns in manufacturing companies. *International Journal of Production Economics*, *210*, 15–26. doi:10.1016/j.ijpe.2019.01.004

González, J., & Granillo, R. (2020). Competencias del ingeniero industrial en la Industria 4.0. *Revista Electrónica de Investigación Educativa*, *22*(e30), 1–14. doi:10.24320/redie.2020.22.e30.2750

Secretaría de Economía de México. (2016). *Crafting the future: a roadmap for Industry 4.0 in Mexico.* Autor.

KEY TERMS AND DEFINITIONS

Big Data: Is a combination of structured, semistructured, and unstructured data collected by organizations that can be mined for information and used in machine learning projects, predictive modeling and other advanced analytics applications.

Fourth Industrial Revolution: Represents a fundamental change in the way we live, work and relate to one another. It is a new chapter in human development, enabled by extraordinary technology advances commensurate with those of the first, second and third industrial revolutions. These advances are merging the physical, digital, and biological worlds in ways that create both huge promise and potential peril. The speed, breadth and depth of this revolution is forcing us to rethink how countries develop, how organisations create value and even what it means to be human.

Industry 4.0: Refers to a new phase in the Industrial Revolution that focuses heavily on interconnectivity, automation, machine learning, and real-time data. Industry 4.0, also sometimes referred to as IIoT or smart manufacturing, marries physical production and operations with smart digital technology, machine learning, and big data to create a more holistic and better connected ecosystem for companies that focus on manufacturing and supply chain management. While every company and organization operating today is different, they all face a common challenge—the need for connectedness and access to real-time insights across processes, partners, products, and people.

Training: The process of learning the skills you need to do a particular job or activity.

Chapter 4
Student Attitude in Mathematics and Science-Related Courses for Digitally Based Society

Salako E. Adekunle
https://orcid.org/0000-0001-9021-9855
Department of Computer Science, FCT College of Education, Nigeria

ABSTRACT

One of the principal objectives of teaching and learning mathematics, science, and technology (MST) is to produce a digitally-based society, and students are pillars to achieve this fundamental objective in this information age. The study investigated students' attitude in MST education for digitally-based society in the South-West region, Nigeria. A sample of 1,080 from the population of 2,700 was used from six private and six public tertiary institutions in the six states that made up the region. A questionnaire with a reliability coefficient of 0.813 was used to gather research data for analysis. Findings revealed that the students in the South-West geopolitical region had a poor attitude in mathematics, chemistry, physics, and computer science, and there was a poor academic performance in MST education. There was a significant difference between public and private tertiary institutions students' attitude towards MST-related courses. Recommendations on periodical workshops and training of teachers on the innovativeness in the teaching of science-based subjects were made.

DOI: 10.4018/978-1-7998-3868-5.ch004

INTRODUCTION

The solutions to the challenges in society require constant but technological approaches. In this information age, Mathematics, Science, and Technology (MST) play a fundamental role in getting work done easily and achieving specific objectives. The teaching and learning of science and technology have brought a great dynamic of innovativeness in the education sector and the world at large. The learning of science and technology has potentials to positively enhance mental capabilities in academic activities for efficient outputs. One of the core objectives of teaching and learning Mathematics, Science, and Technology (MST) at different levels of education is to yield two classes of scholars; the scientists and the engineers who would continue researches that would tackle human challenges in the societies and enhance the economic growth of nations in the world. A digitally-based society is achievable through a functional MST education (MSTE).

A digitally-based society is a community that is driven by modern technologies. Practically, activities are carried out in a digitally-based society using powerful tools of science and technology. For example, business transactions could be performed online in a digitally-based society.

Also, through MSTE, workers that are technological oriented to occupy the workforce are produced. In this information age, the societies need to be technologically driven for effective communication, trading, transportation and governance. The contemporary technology is majorly based on the sciences, with disciplines such as Physics, Chemistry, Biology, Mathematics, Further Mathematics, Computer Science and other sciences (Ajagun, 2016) that deal with the study, understanding, and evaluation of natural phenomena. Effective Mathematics, Science, and Technology (MST) education is paramount in the acquisition of:

1. Knowledge of the complexities of nature and human life.
2. Technological and logical skills.
3. Vocational and technical education for self-reliance.
4. Scientific knowledge in solving educational and societal problems at large.
5. Abilities and competencies for the development of society.

The aforementioned benefits imply that there is a need for a positive attitude in preparing students in the tertiary institutions for occupations in highly scientific and technological societies. This opinion was suggested in John (2018) that students ought to have been adequately equipped with mathematical knowledge and skills for effective learning of science courses. The society is dominated by sectors such as telecommunication firms, engineering companies, health centres, aviation industry, financial institutions, education to mention but a few. Therefore, for high productivity in these sectors, the students need adequate knowledge in Mathematics, Science, and Technology as a foundation for future effectiveness in future services. The question is, what is the level of students' academic performance in Mathematics, Science, and Technology Education for a digitally-based society?

Moreover, quality education, sustainable cities, communities, industry, innovation and infrastructure are among the sustainable development goals (SDGs) adopted by the World General Assembly in September 2015 and to be achieved in 2030 (Peter, Martin, David & Daphne, 2017). Also, World Summit for Sustainable Development (WSSD) suggested that people should be imparted with technological skills to accomplish the three (3) goals of value–orientation, poverty eradication, and employment generation (Ajeyelemi, 2006). Attainment of all these stated objectives rests on the shoulder of students, and the attitude exhibits in MSTE is crucial. The students are the leaders, managers, wealth creators, decision-makers, employers, and educationists of tomorrow. Therefore, the students need to develop the right concepts and attitude in MSTE for technologically-driven societies.

However, Nworgu (2007) reported that the performance of Nigerian students in MST-related subjects is poor. Thus, it would be difficult to achieve scientific and technological advancement in Nigeria (Nworgu, 2007). Evidence from the reports showed that Nigerian students performed poorly in MST selected subjects like Further Mathematics, Mathematics, Biology, Chemistry, and Physics both in external and internal examinations (Chukwuneke & Nwachukwu, 2005; Ajagun, 2016). This was further explained by the West African Examinations report of 2018 that revealed a high failure rate in Mathematics (John, 2018). Wendy (2018) advised students to develop a good attitude and adequate preparation for an excellent performance. Tata, Abba and Abdullahi (2014) revealed that the students in Bauchi State of Nigeria had poor academic performance in Mathematics and factors responsible the poor performance were negative attitude, poor teaching methods and lack of qualified teachers. In addition, Yusuf and Araba (2019) indicated that students in Kwara, Nigeria performed poorly in Mathematics and causes of failure in Mathematics revealed included poor teaching method and students' socio-economic background.

In Hafiz and Hina (2016), it was reported that there was a low students' performance in Mathematics which is similar to the experience in Nigeria. The low performance was attributed to poor students' perception on Mathematics. A poor academic performance in Mathematics was also reported in South Africa by Zingiswa (2019) and causes of poor performance listed included poor attitudes and teachers' training and empowerment.

Teaching methods are the strategies used by the teachers, lecturers and instructors at all levels of education to convey instructions, relevant skills, and knowledge to the learners. The teachers, lecturers and instructors could adopt any suitable teaching approach to positively develop students' interests and learning capabilities in MST-related courses towards the attainment of specific educational objectives. In science-related courses, many teaching methods had been employed. Among these methods were lecture, tutorial, lecture-demonstration, game-based learning, cooperative learning, modelling, simulation, feedback, inquiry-based learning, assignment, experiential learning, class discussion, student-led classroom, inquiry-guided instruction, graphic organizers, field trips, class projects to mention but a few. The teachers' characteristics might include qualifications, teaching experience, attitudes, competencies, approachability, knowledge of subject-matter, communication skills and so on. Since teaching and learning is two-way communication between teachers and students, the teaching strategies, as well as teachers' characteristics, could influence students' attitude in mathematics, science and technology education for digitally-based society.

EMPIRICAL STUDIES

Belbase (2013) presented a relationship between the students' images, anxieties, and attitudes in learning Mathematics. The finding indicated that there was a connection between students' anxieties and attitude in Mathematics. Belbase (2017) failed to examine the level of students' academic performance in MST education. The research only focused on a subject while this present research was on three additional science-related courses and the level of students' academic performance in Mathematics, Chemistry, Physics and Computer Science was investigated.

Metin (2015) investigated students' attitudes towards Mathematics in Denizli, Turkey. In the study, it was discovered that students' attitudes were at the medium level. Furthermore, mothers' profession, fathers' educational level, and gender affected the students' attitudes toward learning Mathematics. However, this present research examined the difference between students' attitude and the teaching methods as against mothers' profession, fathers' educational level, and gender in Metin (2015).

The research of Mislina and Fauziah (2017) was on the attitude of students in Mathematics. The research was conducted in one of the Universities in Malaysia. The finding indicated that a group discussion could assist students to learn and develop a good attitude towards Mathematics. In Mislina and Fauziah (2017), there were no statistical reports on the major teaching methods adopted by the lecturers in Mathematics. The research only focused on a subject while this present research was on four science-related courses, Mathematics, Physics, Chemistry and Computer Science for a digitally-based society. Furthermore, a report on the major teaching methods adopted by the lecturers in MSTE was presented in this current research.

Opara et al. (2017) examined studied the factors that affect students' attitude towards Mathematics in Obio-Akpor, Rivers State, Nigeria. The population was the students in senior secondary schools. It was discovered that self-efficacy, self-concept, and teachers' competency affected students' attitude towards learning Mathematics. The research of Opara et al. (2017) specifically focused on the students' attitude in Mathematics and senior secondary schools. However, this present research covered the students in the Colleges of Education, Polytechnics, and University not senior secondary schools. Also, students' attitude in Mathematics, Chemistry, Physics and Computer Science were investigated as against only Mathematics in Opara et al. (2017). This research was different from Opara et al. (2017) in the area of teaching methods adopted to impart knowledge in Mathematics and other MST-related courses.

Mzomwe, Calkin and Respickius (2019) examined students' attitude and learning Mathematics in Tanzania. It was gathered that students had a less positive attitude as the students were promoted from lower to higher classes. The affecting factors highlighted were teachers' moral/institutional strategies, poor examination strategies, and poor instruction delivery. The research specifically focused on the students' attitude in Mathematics and the College. However, this present research covered the students in the Colleges of Education, Polytechnics, and University to make a general conclusion on students' attitude in Mathematics, Chemistry, Physics and Computer Science.

Research Problem

In Olubusuyi (2003) and John (2018), it was reported that the Nigerian students' academic performance in MST was rated low when compared with other West African countries for nine (9) successive years. Obomanu and Adaramola (2011) highlighted that secondary schools' students did perform poorly at the cognitive, psychomotor, and affective domains levels in Mathematics and other science-related courses. This unimpressive performance in MST and other science-related subjects had prevented majority of the students to gain admission into Colleges of Education, Polytechnics and Universities to study disciplines related to MST. The teaching strategies adopted by teachers and lecturers were among the factors affecting students' performance in science-related courses (Ajagun, 2016). The poor academic performance of students in MST-related courses is a major problem that could endanger technological driven societies. This implies that few graduates would become employable into industries driven by technology. Azuka (2000) pointed out that Mathematics is the pillar of the sciences, and the poor academic performance in this subject would negatively affect the overall academic performance in all other science-related courses. The chapter would investigate the students' attitude in Mathematics and other MST-related courses like Chemistry, Physics and Computer Science for digitally-based society in tertiary institutions of the South-West region of Nigeria.

Research Questions

The research questions were:

1. What are the major causes of poor attitudes towards learning Mathematics and other science-related courses?
2. What are the major teaching methods adopted for teaching Mathematics and other science-related courses?
3. What is the attitude of students towards MST related courses for digitally-based society?
4. What is the level of students' academic performance in MST education?
5. Is there any difference between female and male students' academic performance in MST education?

Hypotheses

The research hypotheses postulated were:

1. There is no significant difference between public and private tertiary institutions students' attitude towards MST education.
2. There is no significant difference between female and male tertiary institutions students' attitude towards MST-related courses.
3. There is no significant difference in the attitude of tertiary institutions' students towards MST-related courses in the states of the region.
4. There is no significant difference in the attitude of tertiary institutions' students towards MST-related courses and the teaching methods.

Objectives of the Study

Specifically, the research attempted to achieve the following objectives:

1. to identify the major causes of poor attitudes towards learning Mathematics and other science-related courses,
2. to establish the major teaching methods adopted for teaching Mathematics and other science-related courses,
3. to find out the attitude of students towards MST-related courses for digitally-based society,
4. to examine the level of students' academic performance in MST education,
5. to ascertain the difference between male and female students' academic performance in MST education,

6. to investigate any significant difference between public and private tertiary institutions students' attitude towards MST education,
7. to determine any significant difference between female and male tertiary institutions students' attitude towards MST-related courses, and
8. to establish any significant difference in the attitude of tertiary institutions' students towards MST-related courses in the states of the region.
9. to pinpoint any significant difference in the attitude of tertiary institutions' students towards MST-related courses and the teaching methods.

METHODOLOGY

A descriptive survey was used to conduct this research. The population was 2700 students of various Universities (first year/100Level) in the South-West geopolitical region of Nigeria with six (6) states, namely Oyo, Ekiti, Lagos, Ondo, Osun, and Ogun. A simple random sampling method (SRSM) was used to select 1080 students from six private and six public institutions based on 15 students per tertiary institution from each state in the geopolitical region.

A questionnaire was designed by the researcher as a research instrument to gather relevant data for statistical analysis and interpretation. The instrument was structured and scored on a 4-point Likert–type scale as; Strongly Agree (4), Agree (3), Disagree (2) and Strongly Disagree (1). The highest scale value was 4 and the lowest scale value was 1. The instrument was validated by experts in Educational Measurement and Evaluation to establish both face and content validity of the instrument. A pilot test was conducted at FCT College of Education, Zuba and University of Abuja to establish the reliability co-efficient and a statistical value of 0.813 was obtained. The administration of the questionnaires for data collection was achieved by the principal researcher and other twelve (12) research assistants under the specific instructions of the principal researcher.

There were eight (8) items on the research instrument that dealt with the students' attitude and MST education. Furthermore, students' academic performance was collected from the academic records provided by the management of the selected tertiary institutions. Based on a structured eight-item to evaluate students' attitude, Equation 1 was used for final interpretation:

$$\frac{(\text{Highest scale value} \times 8) - (\text{Lowest scale value} \times 8)}{\text{Number of levels}} \tag{1}$$

Thus:

$$\frac{(4\times 8)-(1\times 8)}{2}=\frac{32-8}{2}=\frac{24}{2}=12$$

This implied that 32 was the highest obtainable score and 8 was the lowest obtainable score. Thus, a mean between 8 and 20.00 (8+12) was remarked "Poor Attitude" while a mean greater than 20.00 (i.e 20.01) was remarked "Good Attitude".

The research questions were analysed by using mean and standard deviations while t-test and Analysis of Variance (ANOVA) statistical analysis were used to test the null hypotheses postulated for the study at 0.05 level of significance.

RESULTS

The results of the study were presented and discussed as follows:

Research question 1: What are the major causes of poor attitudes towards learning Mathematics and other science-related courses?

Table 1. Analysis of causes of poor attitudes in MST-related courses

Causes	Mean	Std. Deviation
Anxiety	26.24	**7.26**
Peer-group	16.03	**7.71**
Teacher's characteristics	27.42	**7.25**
Lack of learning resources	17.81	**7.23**
Learning environment	13.56	**7.04**
Teaching methods	25.09	**7.32**
Parents' qualifications	14.27	**7.36**
Cluster Mean	**20.06**	

Table 1 shows the mean and standard deviation of causes of poor attitudes towards learning Mathematics and other science-related courses and graphically represented in Figure 1. The mean of each teaching method was examined and it was discovered that anxiety, teacher's characteristics, and teaching methods were causes of poor attitudes towards learning Mathematics and other science-related courses in the selected tertiary institutions.

Figure 1. Bar chart of causes of poor attitudes in MST-related courses

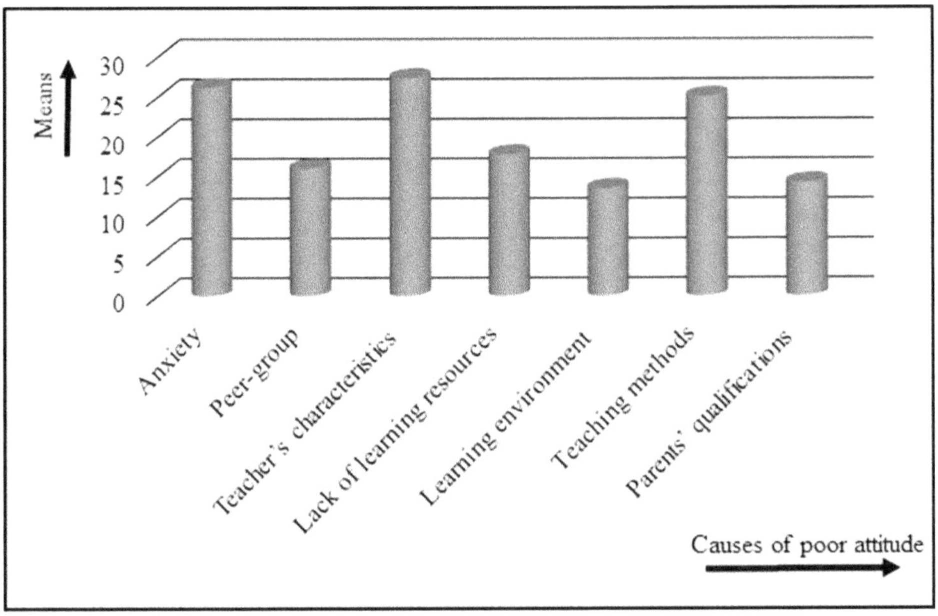

Research question 2: What are the major teaching methods adopted for teaching Mathematics and other science-related courses?

Table 2. Analysis of teaching methods adopted in MST-related courses

Teaching Methods	Mean	Std. Deviation
Lecture	16.07	**7.79**
Inquiry-guided	10.27	**7.05**
Tutorial	9.76	**7.76**
Class discussion	14.88	**7.66**
Field trips	11.03	**7.78**
Assignment	15.20	**7.45**
Simulation	9.67	**6.38**
Cluster Mean	**12.41**	

Table 2 shows the mean and standard deviation of teaching methods adopted in teaching and learning MST-related courses and graphically represented in Figure 1. The mean of each teaching method was examined and it was discovered that lecture, class discussion, and assignment were the major teaching methods adopted for teaching Mathematics and other science-related courses in the selected tertiary institutions.

Figure 2. Bar chart of teaching methods adopted in MST-related courses

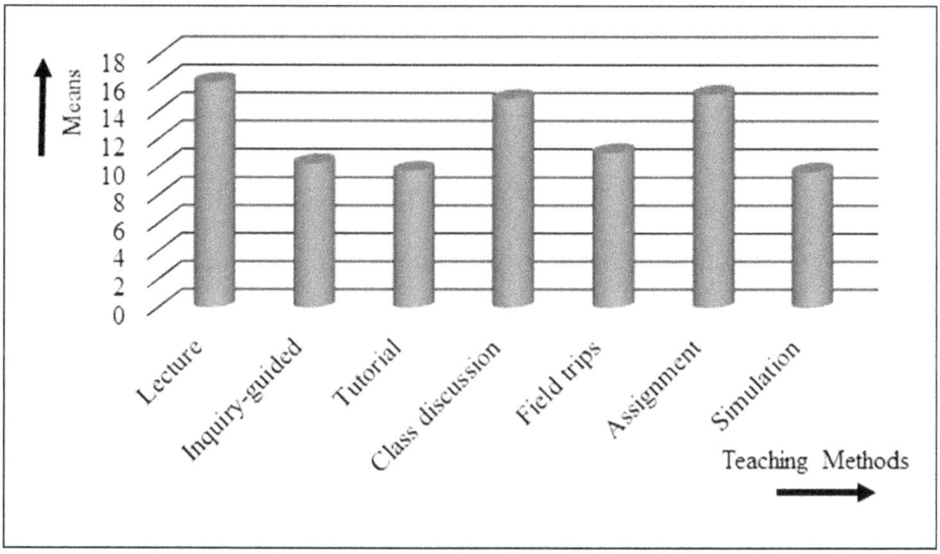

Research question 3: What is the attitude of students towards MST-related courses for digitally-based society?

Table 3. Analysis of students' attitude towards MST-related courses

Subjects	Mean	Std. Deviation
Mathematics	10.80	**6.73**
Chemistry	12.21	**7.34**
Physics	11.89	**7.63**
Computer	11.28	**7.39**
Cluster Mean	**11.55**	

Table 3 shows the mean and standard deviation of students' attitude towards MST-related courses and graphically represented in Figure 2. The statistical means of Mathematics, Chemistry, Physics and Computer Science were 10.80, 12.21,

11.89, and 11.28 respectively. Each mean was lesser than the reference mean of 20.00 which indicated poor attitude. Expressively, the performance of students in Mathematics was poor as it had the lowest mean of 10.80. Additionally, the cluster mean of 11.55 was within a mean range of 8.00 to 20.00. Conclusively, the students in the tertiary institutions of South-West geopolitical region of Nigeria had a poor attitude towards MST-related courses such as Mathematics, Chemistry, Physics and Computer Science.

Figure 3. Bar chart of students' attitude towards MST-related courses

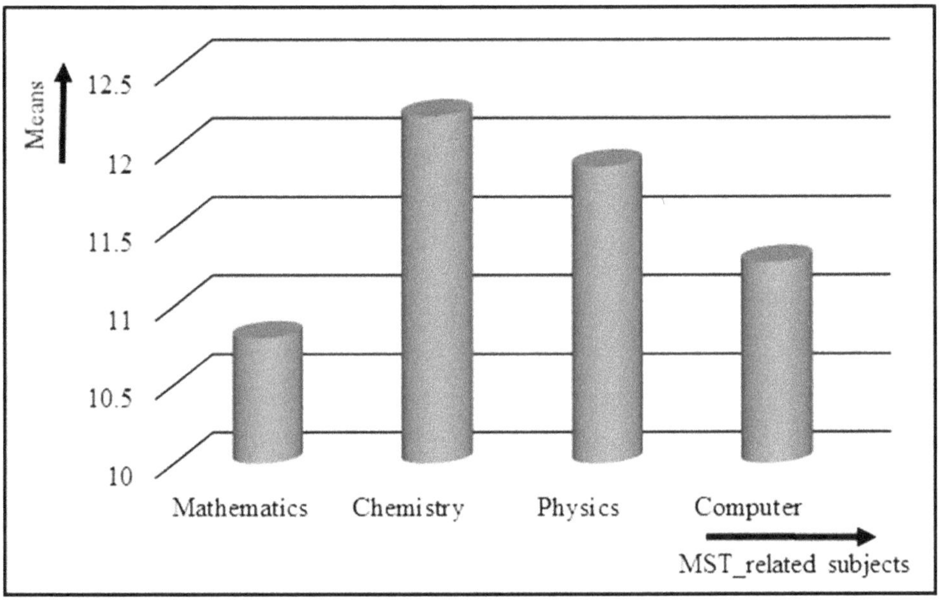

Research question 4: What is the level of students' academic performance in MST education?

Table 4. Students' level of performance in MST education

Performance	N	Minimum	Maximum	Mean	Std. Deviation
Level	1080	2.00	28.00	13.23	**6.80**
Valid N	**1080**				

Table 4 presents an analysis of students' level of academic performance in MST education. To determine the level of students' academic performance in MST-related courses, the scores of students in 2019/2020 session as made available by the examination and academic records of each institution were used. The scores of each student were analysed for presentation and interpretation. The minimum and maximum means were 2.00 and 28.00 respectively. The overall performance was calculated as 13.23, which was within a mean range of 8.00 to 20.00. This implied that the students in the South-West geopolitical region had a poor academic performance in MST education.

Research question 5: Is there any difference between female and male students' academic performance in MST education?

Table 5. Gender analysis of students' performance in MST education

Gender	N	Mean	Std. Deviation	Mean Difference
Male	647	15.48	7.67	
				0.92
Female	433	14.56	7.35	

Table 5 shows gender analysis of students' academic performance in MST education and graphically represented in Figure 4. The mean difference between male and female students' performance was 0.92. This implied that there was a statistical difference between male and female students' academic performance in MST education and male students performed better than their female counterparts.

Testing of Research Hypotheses

The null hypotheses were tested as follow:

***H*$_{01}$:** There is no significant difference between public and private tertiary institutions students' attitude towards MST education.

Table 6 shows a calculated t-value of 12.69 at a tested value of 0.05 level of significance. Since the computed p-value of 0.001 was lesser than the level of significance value of 0.05, this implied that there was a statistically significant difference between public and private tertiary institutions students' attitude towards MST education.

***H*$_{02}$:** There is no significant difference between female and male tertiary institutions students' attitude towards MST-related courses.

Figure 4. Bar chart of students' gender and attitude towards MST education

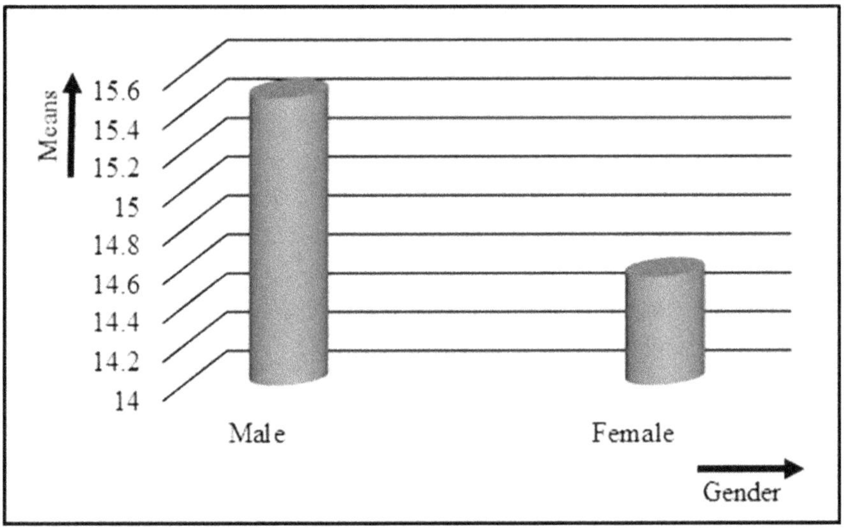

Table 6. T-test analysis of public and private students' attitude towards MST education

Schools	**N $\overline{X}$ SD Df t-cal. Sig (2-tailed)**
Public **Private**	752 9.07 7.06 1078 12.69 0.001 328 14.11 2.04

P (0.05)

Table 7 shows a calculated t-value of 1.970 at a tested value of 0.05 level of significance. Since the computed p-value of 0.088 was greater than the level of significance value of 0.05, this implied that there was no statistically significant difference between female and male tertiary institutions students' attitude towards MST-related courses.

Table 7. T-test analysis of female and male students' attitude towards MST education

Gender	**N $\overline{X}$ SD Df t-cal. Sig (2-tailed)**
Male **Female**	647 15.48 7.67 1078 1.970 **0.088** 433 14.56 7.35

P (0.05)

H_{03}: There is no significant difference in the attitude of tertiary institutions' students towards MST related courses in the states of the region.

Table 8. ANOVA analysis of students' attitude towards MST education

Variable	Sum of Squares	df	Mean Square	F	Sig.
Between Groups	1810.74	5	362.15	6.51	**0.000**
Within Groups	59736.37	1074	55.62		
Total	**61547.11**	**1079**			

P (0.05)

Table 8 shows an F-value of 6.51 at a tested value of 0.05 level of significance. Since the computed p-value of 0.000 was lesser than the level of significance value of 0.05, this implied that there was a statistically significant difference in the attitude of tertiary institutions' students towards MST-related courses based on the states in the geopolitical region.

H_{04}: There is no significant difference in the attitude of tertiary institutions' students towards MST-related courses and the teaching methods.

Table 9. ANOVA analysis of students' attitude and teaching methods

	Sum of Squares	Df	Mean Square	F	Sig.
Between Groups	810.747	6	135.125	2.387	**0.027**
Within Groups	60736.360	1073	56.604		
Total	**61547.107**	**1079**			

P (0.05)

Table 9 shows an F-value of 2.387 at a tested value of 0.05 level of significance and the computed p-value of 0.027 was lesser than the level of significance value of 0.05. Previously, it was emphasised in Table 2 that lecture, class discussion, and assignment were the major teaching methods adopted for teaching Mathematics and other science-related courses. These statistical values of Table 9 reveal that there was a statistically significant difference in the attitude of tertiary institutions' students towards MST-related courses and the teaching methods.

Discussion

The finding of this research indicated that anxiety, teacher's characteristics, and teaching methods were causes of poor attitudes towards learning Mathematics and other science-related courses in the selected tertiary institutions. This result was similar to Belbase (2013) and Opara et al. (2017) that discovered a psychological impact on students' attitudes towards learning Mathematics. In addition, it was revealed that lecture, class discussion, and assignment were the major the teaching methods adopted in teaching Mathematics and other science-related courses in the tertiary institutions of South-West region of Nigeria, this was similar to Obomanu and Adaramola (2011). It was further discovered that the students in the South-West geopolitical region had a poor attitude towards MST related courses such as Mathematics, Chemistry, Physics and Computer Science and this was similar to the findings of Awosiyan (2006) and John (2018). Furthermore, the students in the South-West geopolitical region had a poor academic performance in MST education and this was as well comparable to John (2018) results. It was gathered that there was a statistical difference between male and female students' academic performance in MST education. This finding was relatively similar to Mzomwe, Calkin and Respickius (2019) that revealed the difference between male and female college students' attitude in Mathematics. The male students performed relatively better than the female students in MST education. The testing of the null hypotheses postulated showed that there was a significant difference between private and public tertiary institutions students' attitude towards MST education, there was no significant difference between male and female tertiary institutions students' attitude towards MST-related courses, there was a significant difference in the attitude of tertiary institutions' students' towards MST education based on the states in the geopolitical region, and there was a statistically significant difference in the attitude of tertiary institutions' students towards MST-related courses in the teaching methods. This implied that students' attitude differed with diverse teaching methods adopted by the teachers in the tertiary institutions of the South-West region of Nigeria. This finding was different from Metin (2015) who considered mothers' profession, fathers' educational level, and gender as factors affecting students' attitude in Mathematics only not included sciences courses like Physics, Chemistry and Computer Science. The stated objectives were achieved through the use and analysis of statistical data colleted from the respondents.

Implications on MST-Related Courses

The teachers and other stakeholders in education needed to be equipped with relevant information for the higher academic performance of students in MST education.

This research would assist the lecturers at the tertiary institutions to identify causes of poor attitudes towards MSTE and help lecturers to select appropriate teaching methods that would promote effective teaching and learning of Mathematics and other science-related courses for digitally-based society. Significantly, the information in this chapter would serve as literature for further investigation in MST-related courses.

CONCLUSION

Students' anxiety, teacher's characteristics, and methods of teaching and learning were major causes of poor attitudes towards the attainment of higher performance in Mathematics and other science-related courses in the selected tertiary institutions. The significance of digitally-based society would never be ignored in this dispensation of information driven-world. The present schools' students required relevant skills and knowledge in the field of science and technology for a secured tomorrow against the societal challenges. Regrettably, poor attitude towards science and technology-based subjects would result in poor academic performance in the subjects and finally constitute a challenged society. It was discovered that the students in the South-West geopolitical region had a poor attitude and poor academic performance in MST-related courses. The teachers of MST-related courses needed to have a good knowledge of students' preferred choice of method for effective learning that would provoke digitally-based societies.

RECOMMENDATIONS

The critical findings of this research inspired the following recommendations:

1. Adequate funds should be provided towards the procurement of instructional materials to create a conducive environment for students' participation and understanding of MST-related subjects like Mathematics, Physics, Chemistry, and Computer Science.
2. Scholarship should be awarded to secondary school students for motivation in choosing careers in MST disciplines.
3. Workshops and training should be provided for teachers on the teaching methods that would provoke innovativeness in the teaching of science-related courses.
4. Parents and guardians should provide relevant school materials (books, computer) for their children and wards towards effective participation in the MST discussion.

5. Excursion and field trips should be organized so that students can appreciate the significance of MST in society.
6. Appropriate evaluation by the tertiary institutions' administrators should be provided to observe students' performance in MST subjects and recommendation should be made for a positive attitude.

REFERENCES

Adegboye, S. O. (2013). The senior secondary school science curriculum and candidates' performance. An appraisal of 1st cycle of operation. *Journal of Science Teachers Association of Nigeria. STAN, 28*(1 & 2), 3–12.

Ajagun, G. A. (2016). Towards good performance in science education. *Nigerian Journal of Teacher Education and Teaching, 2*(1), 117–125.

Ajeyelemi. (2006). *New directions in the Nigerian educational system.* Being the 10th annual public lecture of the Uguogie Irowi foundation held on 8th December at NERDC conference center, Agidingbi, Ikeja, Lagos.

Awosiyan, Y. O. (2006). A survey of problems in teaching and learning of mathematics in secondary schools in Damaturu L.G.A., Yobe State. ABACUS. *The Journal of mathematics association of Nigeria, 31*(1), 130 – 141.

Azuka, B. F. (2000). Mathematics in technological development: Focus on the next millennium – implications for secondary education. *The Journal of Mathematics Association of Nigeria., 25*(1), 74–83.

Belbase, S. (2013). Images, anxieties, and attitudes toward mathematics. *International Journal of Education in Mathematics. Science and Technology, 1*(4), 230–237.

Chukwuneke, B. U., & Nwachkwu, C. D. (2005). State and future of education in Nigeria. In J. O. Afe & G. C. Edozie (Eds.), *Books of abstract and lead papers (pp. 58 – 59). Asaba, Nigeria: West and Solomon Publishing Company.*

Hafiz, T. J., & Hina, H. A. (2016). Causes of poor performance in mathematics from teachers, parents and student's perspective. = *American Scientific Research Journal for Engineering, Technology, and Sciences, 15*(1), 122–136.

John, O. N. (2018). WAEC gives reasons there is mass failure in 2018 May/June WASSCE result. *Daily Post.* Retrieved on February 11, 2020 at https://dailypost.ng/2018/07/17/waec-gives-reasons-mass-failure-2018-may-june-wassce-result/

Metin, Y. (2015). High school students' attitudes towards mathematics. *Eurasia Journal of Mathematics, Science & Technology Education, 12*(4), 931-945.

Mislina, A., & Fauziah, K. (2017). Assessing Mathematics' Attitude among Technical University Students. *International Journal of Academic Research in Business & Social Sciences, 7*(4), 103–115.

Mzomwe, Y. M., Calkin, S. M., & Respickius, O. C. (2019). Investigating Students' Attitude towards Learning Mathematics. *International Electronic Journal of Mathematics Education, 14*(1), 207–231.

Nworgu, B. G. (2007). Reforms and quality education in Nigeria. A key note address. In *19th annual conference of the association for promoting quality education in Nigeria.* River State University of Education, Port Harcourt.

Obomanu, B. J., & Adaramola, M. O. (2011). Factors related to underachievement in Mathematics, Science, and Technology education (MSTE) in secondary schools in Rivers State, Nigeria. *World Journal of Education, 1*(1), 102–105. doi:10.5430/wje.v1n1p102

Okebukola, P. (2002). *Beyond the stereotype to new trajectories in science teaching.* Special lecture presented at the 43rd annual conference of STAN and CAMSTE.

Olubusiyi, P. (2003). *Education and manpower. Poor teachers' pay, welfare rebound in students' performance.* Vangard Newspaper.

Opara, I. J., Magnus-Arewa, E. A., & Nwaukwu, C. (2017). Predictors of students' attitude towards mathematics in Obioakpor local government area of rivers state, Nigeria. *International Journal of Education, Learning and Development, 5*(7), 73–85.

Peter, J., Martin, W., David, H., & Daphne, C. (2017). The sustainable development goals and Information and communication technologies. *Indonesian Journal of Sustainability Accounting and Management, 1*(1), 1–15. doi:10.28992/ijsam.v1i1.22

Tata, U. S., Abba, A., & Abdullahi M. S. (2014). The causes of poor performance in mathematics among public senior secondary school students in Azare Metropolis of Bauchi State, Nigeria. *IOSR Journal of Research & Method in Education, 4*(6), 32-40.

Wendy, E. A. (2018). *Drop in WASSCE performance worrying*. Retrieved on February 11, 2020 at https://www.myjoyonline.com/news/2019/March-12th/drop-in-wassce-performance-worrying.php

Yusuf, S., & Araba, H. (2019). Perceived causes of students' failure in mathematics in Kwara state junior secondary schools: Implication for educational managers. *International Journal of Educational Studies in Mathematics*, *6*(1), 19–33.

Zingiswa, J. (2019). *Mathematics Education System in South Africa*. Retrieved at https://www.intechopen.com/books/education-systems-around-the-world/mathematics-education-system-in-south-africa

Chapter 5
Statistics in the Mathematics Curricula in Bulgaria

Marta Zdravkova Sugareva
https://orcid.org/0000-0002-8347-5049
Plovdiv University, Bulgaria

Mariana Koleva Mourgova
University of National and World Economy, Bulgaria

ABSTRACT

Living in the 21st century brings new challenges for the educational system, especially that which concerns education in mathematics, statistics, and technologies. In this chapter, recent problems of teaching statistics in schools in Bulgaria are discussed. The present state of the mathematics curriculum is presented, namely in the section "Probabilities and statistics"; some existing problems are discussed, and recommendations are made for further improvement of the teaching material. The objective of the study is to provide ground for increasing the statistical literacy of the younger generation through a more comprehensive and more up-to-date education in school.

INTRODUCTION

Mathematics and statistics literacy is gaining importance in the contemporary period as smart digital technologies are becoming dominant in developed countries. A whole new concept of education has emerged as a response to the needs of the Fourth Industrial Revolution, characterized by a life-long learning process, more personalized and continuous ("Education 4.0"). Teaching is taking place in various

DOI: 10.4018/978-1-7998-3868-5.ch005

forms in and outside the classroom, using various types of technologies, field works, communication and evaluation methods; it is more adapted to the practical needs, more flexible in time and space.

In this situation, school education has to be updated in order to meet the new challenges; new educational settings and methods have to be developed, enabling the young generation to acquire the knowledge and skills necessary for their performance in a digital-driven world. In this regard, one important step is to update the school mathematics curricula, especially in the section of statistics. It is becoming clear that personal success in many fields of life depends on the quality of mathematical and statistical literacy acquired in school. Moreover, the economic prosperity of every country in the current era will also depend on the literacy of its citizens, combined with their ability to work with modern technologies. Therefore one of the central tasks of school education in the modern era is to create favorable conditions for students and for teachers. This means that a new concept of teaching and learning is evolving, with greater freedom of choice, and more attention to developing a student's capacity for creative thinking and decision making. In this context, education in mathematics and statistics has to be adapted to the new technological environment, as it has to play an important role for the literacy of the younger generation and of the society as a whole.

The purpose of this chapter is to explain the present state of the curricula in mathematics and statistics in secondary schools in Bulgaria and to propose recommendations for its further improvements. The present state of teaching statistics in schools in some other countries – France, Belgium, Italy and the United States – is briefly outlined with respect to the studied subject, in order to provide a comparative perspective to the research. Proposals for future steps in reforming the mathematics school curricula in Bulgaria are presented, namely for the section on Probabilities and statistics.

BACKGROUND

Why is Statistical Literacy Important?

Statistical literacy is an integral part of overall literacy that has to be attained in school before adulthood, along with computer literacy, scientific literacy, cultural literacy, visual literacy, quantitative literacy (Watson& Callingham, 2003). Alongside those other types of literacy, statistical literacy has to play a crucial role for the proper functioning of civil society.

When discussing issues of literacy and illiteracy, the skills most often referred to are those associated with using written language. Less often is illiteracy associated with skills related to understanding numbers, interpreting them correctly, and performing actions with them. However, in recent years the terms "mathematical literacy" and "statistical literacy" are gaining popularity. (Callingham & Watson, 2020).

Contemporary society requires that a certain level of mathematical and statistical literacy be attained by every citizen. According to the UNESCO definition, "literacy is the ability to identify, understand, interpret, create, communicate and compute, using printed and written materials associated with varying contexts." (Montoya, 2018).

Statistical literacy can also be portrayed as "the ability to interpret, critically evaluate, and communicate about statistical information and messages" (Gal, 2002). It guarantees that every citizen can perceive and correctly understand the digital information from the surrounding world, and can successfully operate as an equal member of society. Citizens who are not able to interpret correctly the statistical data they encounter in everyday life, in the media, in the political sphere, can be easily manipulated, and this is threatening the very foundations of civil society.

Statistical data are sometimes suspected to be "fake", aiming to lie and manipulate public opinion. The lack of knowledge and information about the nature of some official statistical data may provoke such suspicion and negative attitudes to statistics in general. Statistical illiteracy, or the lack of sufficient statistical knowledge in the general public, are contributing factors for such attitudes. In the media, we often witness serious errors in the presented numbers and graphs and even more so in their interpretation. Sometime it is difficult to distinguish whether the errors are made intentionally or just as a consequence of incompetence.

That is why statistical literacy is essential in the mass communications of contemporary society, as well as in the everyday communications of every citizen.

A number of definitions of statistical literacy are known in the literature. For example, according to the W. M. Keck Statistical Literacy Project (cited by Schield, 2011) statistical literacy is a critical way of thinking about numbers and statistics used as evidence in disputes; the ability to read and interpret numbers in statements, surveys, tables and graphs; studying how statistical relationships can be used as evidence of causal relationships.

Earlier, Schield (1998) defined statistical literacy as the ability to use statistics and statistical dependencies as evidence when arguing a particular thesis. In fact, reading and understanding numbers often implies statistical literacy, especially when these numbers are representing the results of a statistical "experiment".

Such experiments include sociological and marketing surveys, and medical and pharmaceutical tests designed to justify the use of a new treatment method. Results of such experiments are often shown in the media in order to prove or reinforce a certain thesis. In such cases, the statistical literacy of journalists and commentators proves to be important or even decisive for the impact of the numbers on the public.

It should be added that information including numbers can have a greater impact on the audiences compared to just verbal texts.

False or deviated statistics can sometimes be used in order to manipulate the audience. The ways of using statistics for manipulative purposes are so many and so different that there are already hundreds of books and papers published on this phenomenon in the world. The issue of manipulations is too large, and goes beyond the purpose of this chapter, therefore it will not be discussed here in detail. However, the seriousness of this problem is another reason why statistical literacy needs to be raised in our country. One of the main paths for doing this is undoubtedly through the improvement of mathematical and statistical education in school (Sugareva & Mourgova, 2017).

The development of statistical literacy can increase the abilities of a person to distinguish between true and false information. Therefore mathematics and statistics education has to be updated to a new forms, new teaching methods and techniques, a new school environment, but also a new curriculum. The authors believe that the teaching material in statistics has to play a crucial role as an integral part of mathematics curriculum, providing practicality and effective links to all the other sciences taught in school.

A very useful book on the subject was published in 2011 by Carmen Batanero, Gail Burill and Chris Reading. (Batanero et al., 2011). The book is an overview of a common project on the challenges for teaching statistics in school mathematics and preparing mathematics teachers. The aim was "to reflect on the teaching of statistics in school mathematics and on the training of those teachers who are responsible for this teaching, under the auspices of the International Commission on Mathematical Instruction (ICMI) and the International Association for Statistical Education (IASE)."

The authors point out that the results of their work "have contributed to raising awareness of the need for increased statistical content at school levels to improve statistical literacy in young students around the world as well as awareness of the related challenges in training and supporting mathematics teachers who teach statistics."

The results of this study show that "even when many prospective secondary teachers have a major in mathematics, few of them have received specific preparation in designing sample collections or experiments, analysing data from real applications or using statistical software (Batanero & Dнaz, 2010). These teachers also need education in the pedagogical knowledge related to teaching statistics ..." (Batanero et al., 2011, p. 6). The authors point out that "the situation is even more challenging for primary teachers, ... teaching statistics to these children needs different approaches, tasks and methods than teaching statistics in secondary or high school."(Batanero et al., op. cit.)

There is no evidence that Bulgarian specialists have been participating in this common project.

Mathematics and Statistics Curricula in the Frame of Education 4.0

Education 4.0 is a new educational system of ideas and practices, designed to transform primary and secondary education systems in order to prepare the younger generation for the new digital challenges of the information age, now sometimes referred to as "Industry 4.0", a reference to the current age being the Fourth industrial revolution.

Education 4.0 was recognized as a response to Industry 4.0, greatly increasing the use of Internet technologies and crosscommunication tools. ...Education 4.0 is developed for Industry 4.0 and prepares qualified professionals for a very global and digital work environment. (Sharma, 2019)

The new technologies in mathematics and statistics education allow not only digital methods of teaching, but also the introduction of cyber systems for analyzing statistical data in the classroom. Before doing this, it is important to note that students need to have a certain level of statistical literacy, i. e. to master basic statistical notions and methods. Basic statistical literacy is a necessary prerequisite for the correct interpretation of empirical data in any higher steps of data analysis with the use of modern technologies. Therefore, statistical education has to start in elementary school.

During the last decades, technology has become a major part of everyday life, including the schools. School courses in most subjects use internet sources, and various types of teaching technologies in order to facilitate and to improve the quality of the teacher's performance. This is especially true for courses of mathematics and statistics. The positive impact of advanced technologies on teaching and learning statistics can be revealed in various aspects. They allow gathering and processing of huge amounts of statistical data, coming from various sources, and facilitate their processing. Communication between teachers, between the teacher and the students, and between the students is performed in new and challenging forms. Computation procedures are getting much easier and faster, which gives the educators the opportunity to emphasize other issues of statistical analysis, such as explaining statistical methods, identifying regularities and trends, and interpretation of the results.

Another direction of the use of digital technologies can be detected in the increased possibilities for graphical representations and data visualization (Tishkovskaya & Lancaster, 2012). The transition between tables and graphs is a fascinating procedure of data analysis which intrigues students and facilitates their perception of some important aspects of statistical thinking. Graphical representations are also useful for understanding the structure of complex data, and they can be used to help young students formulate conclusions based on data analysis.

Statistical data are becoming more and more accessible from various sources through the internet and different kinds of digital technologies. The education system should incorporate these possibilities in the teaching process of statistics. In school the students should acquire the skills and the abilities to select the data necessary for solving a given problem, as well as to process and analyze them.

In order to meet the challenges of Education 4.0, the school curricula in mathematics and statistics should be transformed accordingly. Tishkovskaya & Lancaster (op. cit., citing Garfield & Ahlgren, 1988; Garfield, 1995; Moore, 1997) also point out that the necessity to reform statistics instruction and curriculum "based on strong synergies among content, pedagogy, and technology".

STATISTICAL EDUCATION IN BULGARIA

According to results obtained in 2018 by the Programme for International Student Assessment (PISA) conducted among 15-year-old students in 72 countries, Bulgaria has a mean score in mathematics of 436. The OECD average is 489 (Schleicher, 2019). The scores range between 591 in China and 325 in the Dominican Republic. The highest score for Bulgaria was observed in 2015; it was then 441. This score is higher than the score from 2018, but it is lower than the OECD average for the same year, 2015. In spite of some high results of Bulgarian students in international mathematics competitions during recent years, the average level, as demonstrated by the PISA results is not satisfactory; it is below the OECD average.

A question arises: what steps would be appropriate to improve the teaching of mathematics and statistics in schools in Bulgaria?

In Bulgaria, curricula in schools across all disciplines and in all classes, including mathematics, change very often. Changing the curricula until now usually consisted of adding more teaching material, which increased the total burden on students without achieving necessarily positive results. There is an increasing demand for optimizing the work load of students in such a way that by the end of gymnasium they would have the proper knowledge and skills for a contemporary citizen and a solid basis for different higher studies at the university level. To this end, efforts have to be made to bring the mathematics curricula to a state which would be more accessible for the students and more efficient at the same time.

Methodology

The main research methods used in the chapter are qualitative and descriptive; observation and research of documents – namely, of the existing textbooks and curricula in Bulgaria and in some other countries for all school grades. This study is undertaken as a case study. The authors have examined the mathematics textbooks and curricula in Bulgaria, mainly in the sections concerning statistics and probability. Examples from other countries (France, Belgium, Italy and the United States) are provided in order to illustrate some possibilities to improve the Bulgarian mathematics curricula.

Statistics in the Mathematics Curriculum in Bulgaria

The pre-university educational system in Bulgaria consists of two levels, basic and secondary. Basic education includes primary school (grades 1-4) and junior high school (grades 5-7). Grades from the Certificate for Basic Education are used as admission criteria for high schools. Secondary education (grades 8-12) comprises high schools and vocational schools. Students can enroll in high school after the successful completion of grades 7 or 8. The admission criteria are based upon grades in Bulgarian language and literature and in mathematics or other subjects in junior high school.

Training in statistics as part of the mathematics curriculum has been introduced in Bulgaria in recent years. In the grades 5 to 7 the respective section is labelled as "Elements of Probabilities and statistics", and in the upper grades as "Probabilities and statistics".

While for other parts of the mathematics curricula, the teaching materials have remained essentially the same, the sections concerning probabilities and statistics have been subjected to the largest changes in the last 2-3 years. New heavy teaching material was added into some grades, but students were not properly prepared with the necessary knowledge from previous grades in order to properly understand it. Observations of the performance of secondary school graduates in higher education are demonstrating that many of them lack a basic knowledge of statistics and mathematics necessary to master economic disciplines.

Many of them demonstrate a serious lack of basic knowledge in mathematics, especially in the knowledge of probabilities and statistics. Taking the statistics exam has become a huge challenge for students in economic disciplines. A large number of students at the higher educational level experience difficulties in dealing with statistical data and in applying statistical methods in their research. At the same time statistical competence is becoming more important for contemporary sciences, including economic studies, medical research, and biology.

This situation calls for research of the school curriculum in mathematics, specifically in the sections of probabilities and statistics. The present chapter is based on researching the existing officially published curriculum in mathematics and on the textbooks for all grades in Bulgaria. The research demonstrates that teaching statistics in the secondary school in Bulgaria suffers some serious inconsistencies (Sugareva & Mourgova, 2018).

Teaching material is not distributed in a gradual way from the 5th to the 12th grade; there is no clear scheme for upgrading the knowledge from previous years into each consecutive year. Some basic principles of statistical methods and measures are lacking or are only partially explained. Too much attention is paid to graphical representations, while the essence of statistics (gathering and processing statistical data in order to reveal regularities) is insufficiently explained. This is especially true for the lower grades where the basics of the statistical approach and some of the basic notions of statistics and probabilities should be explained to the students in order to provoke their curiosity and to stimulate their natural need for scientific research.

In this chapter, we focus on curricula in mathematics and statistics in both elementary and secondary educational levels.

a. Curricula in Mathematics and Statistics for Primary and Basic Educational Levels

In the first four years of school (primary school), the curricula in mathematics does not contain a section on "Probabilities and statistics". This section appears for the first time in 5th grade (labeled as "Elements of probabilities and statistics"). Nevertheless from 2nd grade, children start to "collect data from various sources in order to compose numeric expressions and text tasks" (Ministry of Education and Science of Bulgaria, 2016a).

In 3rd grade, students learn how to recognize half, third, quarter and tenth as parts of a whole. They use models with numerical expressions to describe situations involving the relationships: "with ... more than ...", "with ... less than ...", "... times more than ..." and "... times less than ..." (Ministry of Education and Science of Bulgaria, 2017a).

In 4th grade, students study geometric figures (circle, radius, angle measurement; area of a rectangle); and spatial bodies (cube, rectangular parallelepiped, ball, cylinder, cone, pyramid). In the section "Modeling" they perform the arithmetic action of dividing multi-digit numbers by a single-digit number. In this grade it would be possible, and desirable, to introduce to them the idea of "part of a whole", and to recall the knowledge from the 3rd grade about half, third, quarter and tenth as parts of a whole.

Helen Tsakiridou and Eleni Vavyla have studied the possibilities of teaching probability concepts in primary school. (Tsakiridou & Vavyla, 2015). They have come to the conclusion that "the majority of students was able to recognize different events and categorize them depending on their likelihood." Further investigations would reveal the great possibilities existing in the primary school grades to teach statistical and probability notions to the children that would benefit them in secondary school grades.

The notion "part of a whole" is studied in detail in the 5th grade, but the emphasis is on numeric calculations and not so much as a step for understanding statistical distribution. As we mentioned above, in this grade for the first time the curriculum in mathematics contains a section on "Probabilities and statistics". As a result of the training in 5th grade, the students are expected to read and interpret information presented in text, in graphs, in tables or in diagrams; to compare information on linked data sets; and to use graphical and tabular data information to answer questions that are indirectly related to the table or graph. The expected results for the student are: to know the concept of percentage; to be able to calculate a share (percent) of a number; to calculate the number when a share (percentage) is given; to be able to represent the ratio of two numbers (not necessarily representing a part and the whole) as in percent; to be able to solve basic problems dealing with share (part of the whole); and to be able to solve simple problems in economics and finance (Ministry of Education and Science of Bulgaria, 2016d).

It would be desirable that the children learn more about the notions "part of a whole" and "percentage" in the previous grades by simple examples, taken from their surrounding world. In this way, they will be prepared for the teaching material studied in 5th grade. In 5th grade mathematics is taught by a new teacher, as all subjects from 5th grade on are taught by specialized teachers. In every upper grade, new notions (such as: proportion, frequencies, and probabilities) should be introduced gradually, based on the knowledge of those basic concepts that students already have. This is especially important for the transition between the 5th and the 6th grades.

In 5th grade it would also be desirable to consolidate the knowledge about share and percentage, and to start explaining the notions of statistical distribution and probability. It is also necessary to elucidate the notion of linked data sets when introducing such basic statistical concepts as universe as a set of units, characteristics of the units, subsets, share (proportion) and quantitative and qualitative characteristics. It is also desirable to introduce the mean, the median, and the mode.

In 6th grade, the section "Elements of probabilities and statistics" becomes much more developed and includes much more teaching material: sets and operations with them; graphic representation of sets; random phenomenon; probability of a random phenomenon as a ratio of possibilities; data description; arithmetic mean; organizing and presenting the data.

After the training, students are expected to be able to find subsets of a set and the intersection and union of sets; to know the concept of random phenomenon at the simplest level (coin, dice); to know how to calculate the probability of a random phenomenon at the simplest level; to be able to find the arithmetic mean and use it to interpret data; to be able to compare and collate different graphical or tabular representations of the same data; to be able to organize and present empirical data in tables and rectangular diagrams (Ministry of Education and Science of Bulgaria, 2016c).

When comparing the new teaching material in the curriculum for 6th grade with that of the previous grades, it becomes clear that in 6th grade there is an accent on probabilistic knowledge and much less new knowledge on descriptive statistics. Therefore it would be appropriate to provide the students some clarifications about the probabilistic approach as such.

In this grade it would be appropriate to teach frequency (absolute, relative, and cumulative) and frequency distribution. All these notions should be learned through extensive examples, involving textual tasks, statistical tables, and graphical representations of the data.

The new statistical knowledge in 6th grade can be best comprehended and consolidated through cross-curricular links with the other parts of mathematics as well as with the other subjects they study in school: Geography and economics, Man and nature, Information technologies, and Technologies and entrepreneurship.

In 7th grade (the last year before high school), the curriculum in the section "Elements of probabilities and statistics" includes the organization and presentation of data, the construction and interpretation of pie charts, and probability of events. The knowledge and skills expected as a result of the training include the ability to organize and present data in tables and in rectangular diagrams (without histograms), to use data sets in order to answer questions, to solve problems in pie charts and bar charts, and to evaluate the chances (probabilities) of a random outcome.

Students are expected to be able to construct and interpret pie charts, to be able to assess the probability of random outcomes, and to use the chance (probability) of a particular outcome in a given trial (Ministry of Education and Science of Bulgaria, 2016b).

The authors recommend that by the end of 7th grade the students should be familiar with the following notions of statistics: stochastic event, trial, outcome of a trial. They should have a clear idea about probabilistic thinking, and they should be able to distinguish between deterministic and probabilistic events. They also should be able to comment, with examples from real life, on such probabilistic notions as randomness and uncertainty, stochastic experiment, individual trial, and predicting the outcome of an individual trial.

The notion of descriptive statistics from the 6th grade—frequency (absolute, relative, and cumulative), and frequency distribution—should be recalled as well, and their knowledge should be consolidated by suitable examples and tests. In addition, they should practice the calculation of arithmetic mean, mode and median.

They should be able to create statistical tables with frequencies and to calculate various measures from a table of a frequency distribution; to be familiar with the notion of variability and the range; and to be able to identify some basic features of a statistical distribution. The students should be able to distinguish between qualitative and quantitative data and should know how to analyze the respective frequency distributions. The students also should be able to create a line plot from data given in table or in text and to create and interpret data in a histogram and in a scatter plot from numerical data. They should be able as well to choose the best type of graph when asked to represent a given set of data.

b. Curricula in Mathematics and Statistics for the Secondary School Level

In 8th grade, some elements of combinatorics are introduced in the section "Probabilities and statistics", namely permutations, variations and combinations. The expected knowledge and skills for the students from the 8th grade are: to identify and to calculate combinatorial compounds without repetition; to apply basic rules for adding and multiplying in combinatorics; to model with permutations, variations and combinations. Basic combinatorial concepts to be studied in the 8th grade are multiplying and adding possibilities and applying the basic rules for adding and multiplying in combinatorics (Ministry of Education and Science of Bulgaria, 2016d).

The authors suggest that in 8th grade the teaching material should be focused also on the following topics of probabilities and statistics: the arithmetic mean (weighted or unweighted, calculated according to the available data), quartiles, probability distribution, and the normal curve.

In 9th grade the section "Elements of probability and statistics" includes the following knowledge and skills: classic probability, probability of the sum of incompatible events, likelihood of an opposite event, intersection and cross-section of events, and probability of the sum of incompatible events.

As a result of the training, the students should be able to read and interpret information presented in graphs, tables or charts; to calculate classical probability using formulas for permutations, variations and combinations without repetition; and to calculate the classical probability of the sum of compatible and incompatible events. (Ministry of Education and Science of Bulgaria, 2016d).

The authors suggest that in 9th grade teaching material on time series should be introduced. It is recommended that curriculum includes the following statistical notions: trend, random fluctuations, smoothing. The student should be able to identify trends with scatter plots and to make predictions with scatter plots. The ratios, rates, and proportions should be introduced as well. The student should be able to estimate percents from numbers, to compare percents in cunderstand ratios, rates and proportions, to identify equivalent ratios, rates, to compare ratios, to estimate population size using proportions, to find the constant of proportionality from a table, to identify proportional relationships in graphs, to graph proportional relationships, to interpret graphs of proportional relationships. It is desirable the students to be able to complex tables, and to find the . They should be able to create a frequency table from data and to interpret it using absolute and relative frequencies.

Extensive examples and tests should be provided in order to support the teaching process.

The curricula in 10th grade includes the following knowledge: population and sample; descriptive statistics such as mode, median, arithmetic mean, range and quartiles; ordered samples and statistical series. As a result of the training, the student is expected to know the concepts of population and sample and to be able to find the central tendency in data: mode, median, arithmetic mean (Ministry of Education and Science of Bulgaria, 2016d).

In addition to this, the authors recommend that some element of inference statistics be included in the curriculum. The following notions should be taught in this section: variance and standard deviation, sample size, standard deviation of sample observations, standard error, confidence interval, and level of significance. They also recommend that the concepts of representative, random and biased sample be introduced.

In 11th grade, the curricula includes knowledge and skills of the mechanisms for presenting statistical data and calculating basic statistics from data of various sources. This includes the ability to work with authentic data related to sociology, economics, meteorology and laboratory trials; the ability to use a calculator or a computer when calculating on specific topics; the ability to formulate data related tasks; collecting, organizing and presenting relevant data in response to formulated tasks, and applying knowledge in real situations. Students are expected to be able to register, process and present information by various means; to be able to analyze statistics presented in various ways; and to be able to use a proper measure of central tendency for analyzing and presenting data. The possibility of cross-domain links is also mentioned, involving the curricula in civil education, physics, chemistry, biology, medicine, agriculture, ecology, finance, and sociology (Ministry of Education and Science of Bulgaria, 2016d).

The authors recommend that the following be added to complete the curriculum for the 11th grade:

- reading and analyzing statistical tables published by the National statistical office.
- calculating aggregate statistical measures from data in tables and graphs.
- using their knowledge of mathematical functions for analyzing trends in tables and graphs.

In 12th grade, but only for students with a major in mathematics (called "Mathematical schools" in Bulgaria), the curriculum includes: population and sample, statistical frequency, mean, standard deviation, dispersion, and normal curve distribution. It is expected that students should be able to formulate data related tasks, to work with authentic data, to collect, organize and present relevant data in response to formulated tasks, and to apply learned knowledge in order to assess real situations. Possibility of cross-domain curricular links is also mentioned (Ministry of Education and Science of Bulgaria, 2016d).

The authors also recommend that students in this grade also study the complex indicators composed of an "intensive factor" and weights (such as mortality rate, fertility rate per 1000 population). In this way they will recall the formulae for calculating a weighted arithmetic mean, and at the same time they will be able to correctly interpret rates. They should study also in more detail both theoretical probability distribution and frequency (empirical) distributions, and how model distributions can be used to "fit" the data. They should also be acquainted with the basic principles and notions of statistical inference and sample surveys, such as stochastic error, confidence interval, sample size, variance and the relations between them.

In the following table, the authors' suggestions for the teaching material to be included in the curricula in each class in the section "Probabilities and statistics" are summarized.

Statistics in the School Curricula of France, Belgium, Italy, England and the USA

In France and in Belgium, even before entering secondary school (Lyceum), that is, during the years of compulsory education (up to age 16), students acquire some of the basics of statistics. (Dejean, 2018; J.-P. Haure, F. Lavandier, 2018). At this stage of education, the teaching material which prepares for understanding the statistical notations and principles takes up about 1/3 of the teaching material in mathematics, while the notion of statistics does not appear in the lower grades. (Grades in France are counted backwards – the children begin at upper grades, while by the end of high school they are in first grade and finally in the "terminal" grade).

Table 1. Proposals for curriculum of the "Probabilities and statistics" section in the Mathematics curricula in every grade

Grades	Competencies as expected results from training	New notions
5th	• To know the concept of percentage; to calculate a share (percent) of a number; to calculate the number when a share (percentage) is given; to represent the ratio of two numbers as a percent; to solve basic problems dealing with share (part of the whole). • To make graphical representations of share, percentage, and proportions. To interpret graphical representations.	• Share • Percentage • Proportion • Pie charts
6th	• To know sets and operations with them; graphic representation of sets; random phenomenon; probability of a random phenomenon as a ratio of possibilities; data description; arithmetic mean; organizing and presenting the data. • To find the arithmetic mean and use it to interpret data; to compare and collate different graphical or tabular representations of the same data; to organize and present empirical data in tables and rectangular diagrams.	• Set of units (elements). Statistical characteristics of the units. Statistical data. Frequency. Absolute and relative frequency. Frequency distribution. Table of frequency distribution. • Randomness. Random phenomenon. Trial. Expectation of an outcome. Probability of a random phenomenon. • Arithmetic mean.
7th	• To organize and present data in tables and diagrams, to construct and interpret pie charts; to evaluate the chances (probabilities) of a random outcome; to calculate probability of events; to use data sets in order to answer questions, to solve problems using tables and graphs.	• Stochastic event; deterministic and probabilistic events. • Uncertainty, stochastic experiment, individual trial, outcome of an individual trial.
8th	• To construct an empirical statistical distribution using empirical data. To analyze an empirical statistical distribution. To calculate the arithmetic mean, mode and median (weighted and/or unweighted, according to the data presentation.) To construct a scale in a situation where the variable is continuous. To calculate the arithmetic mean from a table using the middle of the intervals. • To distinguish between empirical and probabily distribution • To identify and to calculate combinatorial compounds without repetition; to apply basic rules for adding and multiplying in combinatorics; to model with permutations, variations and combinations. To multiply and add possibilities and apply the basic rules for adding and multiplying in combinatorics.	• Variable. Discrete and continuous variables. Scale. Weights. (Recall the frequencies of a frequency distribution.) • Empirical distribution. Unweighted and weighted arithmetic mean. Variance. Quartiles. (Recall the median.) • Probability distribution. The normal curve. • Dependency. Independent events. Compatible and incompatible events. Impossible and certain events. Probabilities. • Intersections. • Basic rules for adding and multiplying in combinatorics
9th	• To read and interpret information presented in graphs, tables or charts; to calculate classical probability using formulas for permutations, variations and combinations without repetition; to calculate the classical probability of the sum of incompatible events. To calculate conditional probabilities. • To identify trends from scatter plots and to make predictions with scatter plots. • To estimate percentage from numbers, to compare percentages, to understand ratios, rates and proportions. To identify equivalent ratios, to compare ratios, to comment on the population structure using proportions, To find the constant of proportionality from a table, to identify proportional relationships in graphs, to graph and to interpret proportional relationships. To convert between percentages and fractions, to calculate ratios, to construct complex tables, and to find the rate of a change (in %). To create a frequency table from data and to interpret it using absolute and relative frequencies. • To construct graph from time-series, and to identify the trend.	• Classical probability, probability of the sum of incompatible events, likelihood of an opposite event, intersection and cross-section of events, probability of the sum of incompatible events. • Conditional probability. • Ratios, rates, and proportions. Equivalent ratios. Rates. Population structure. Proportionality. • Time series, trend, random fluctuations, smoothing.
10th	• To calculate measures of dispersion: range and quartiles; interquartile range standard deviation and variance. • To define outliers. To comment on a table and/or graph of an empirical statistical distribution. • To know the concepts of population and sample and to find the central tendency in data: mode, median, arithmetic mean – using a sample.	• Dispersion. Range, outliers, quartiles, deciles, interquartile range. Standard deviation and variance. • Population and sample. Sampling procedures. Representative, random and biased sample. • Estimating the value of a parameter based on measured empirical data from a sample. Stochastic error. Confidence interval.
11th	• To present statistical data and calculate basics statistics from data of various sources. To work with authentic data related to sociology, economics, meteorology and laboratory trials; using software for calculating specific statistics. To formulate data related tasks: collecting, organizing and presenting relevant data in response to formulated tasks, and applying statistical methods in real situations. To analyze statistics presented in various ways; to use a proper measure of central tendency for analyzing and presenting data. To use statistics in other subjects, such as: physics, chemistry, biology, medicine, agriculture, ecology, finance, and sociology.	• Data sources • Data collection and processing
12th	• To model the data • To know in more detail both theoretical probability distribution and frequency (empirical) distributions, and how model distributions can be used to "fit" the data. • To know the basic principles of statistical inference: stochastic error, confidence interval, sample size, variance and the relations between them. • To understand complex indicators containing an "intensive factor" and weights (such as the population age distribution for calculating mortality rate and fertility rate per 1000 population.)	• Adequacy of the model. • Statistical inference • Confidence interval • Intensive factor, extensive factor.

In 6th grade in France (corresponding to 6th grade in Bulgaria), there is a section in the mathematical curriculum called: "Organization and management of data", which includes the following topics: proportionality, tables and graphs.

In the 5th grade (7th grade in Bulgaria), the section "Organization and management of data" contains teaching material on proportionality. Another section "Statistics" includes: a glossary of statistical concepts; grouping data by the values of one characteristic; frequency calculation; creating a chart.

The 4th grade (8th grade in Bulgaria) includes proportionality, measuring scale, speed, percentages, and the mean — simple arithmetic mean and weighted arithmetic mean.

In 3rd grade in France (9th grade in Bulgaria) the following concepts of statistics are studied: totality (universe), trait (feature, characteristic); total number (of units in the universe), range; average (arithmetic mean); weights; weighted arithmetic mean; cumulative numbers; frequency; the median. This material is taught slowly and with many examples. Quartiles are also studied in the same year. The median and quartiles show the students the "structure of the aggregate"; it helps them to understand (at a later stage of education) the concept of "dispersion" – here measured by the interquartile range. The concept of "probability" is discussed.

In France, the 2nd and 1st and terminal grades (10th, 11th and 12th grades in Bulgaria) constitute what's called the "lyceum". The material required to prepare for the baccalaureate exam (matriculation in Bulgaria) include: average, variance, standard deviation; median, quartiles, distance between quarters; area of intersection of two sets; proportions, percentages and probabilities; an elementary event; random experiment, equal probable events, probability of occurrence of an event, counter-event, probability of occurrence of the opposite event; conditional probabilities; random variable, mathematical expectation, variance and standard deviation; calculating the probability of an event occurring; determining the law of probability distribution of a random variable; probability; binomial distribution; Bernoulli's experience and Bernoulli's scheme; sample and confidence interval; calculation of the binomial coefficients; normal distribution; accepting or rejecting at 95% probability; the influence of the structure on complex indicators including an intensive and extensive (structural) component. (Sugareva & Mourgova, 2017).

In Belgium, the grades are numbered the same as in Bulgaria. In 5th grade, students study the following material (preparation for the section "Statistics and probabilities" in the upper grades): symmetry and axis of symmetry; ratio (relative numbers), relative share (fraction) and percentage; transition from relative share to percentage and vice versa; measuring, weighing (gross, net, tare); tables and graphs - switching from tables to graphs, reading charts (bar, dot, line and pie charts), reading tables; a pie chart and its relationship to a table and to a percentage distribution; frequencies distribution table - in absolute numbers and in percentages

- with examples. A linkage is made with the teaching material on surface and areas. Students are introduced to the concept of "scale" when drawing a map chart used to represent geographical differences. The arithmetic mean and the median values are studied (with numbers and diagram examples). The following economic concepts are introduced: profit, capital, interest and interest rate.

In 6th grade, the students study relative shares, proportionality, data, and graphical images. For relative shares, they study the relative share as a fraction (ordinary and decimal); dividing the whole into equal parts; a part of a number and percent; interest rate and the interest accrual calculated each year, with many examples. Under proportionality, students study the coefficient of proportionality, how to recognize a situation of proportionality, how to use proportionality (accompanied by many examples), and calculation of percentages. Students learn how to organize and present data, how to read tables and charts and how to organize data into a table —by rows, by columns, including in a two-dimensional frequency table (two-dimensional distribution table). Finally, students learn about bar charts and Cartesian graphs; in the latter, they learn how to depict the dynamics of one dimension depending on the values of another dimension.

In 7th grade, the students study proportionality, percentages, speed, and symmetry (with exercises). In the statistics section, they study bar charts, pie charts, and frequencies.

In the 8th grade, they study proportionality and percentages, and in the statistics section, they study data processing, arithmetic mean (simple and weighted), and solve problems with "weights".

In the 9th grade, they study linear function and its parameters, coefficient and intercept, graphical representation, random experiment, random event, elementary event, outcome, compound event (event of a random experiment), opposite event; impossible event; certain, even, incompatible events, probability of an event, probability of a certain event, of an impossible event, probability of occurrence of at least one of two incompatible events (event A or event B).

In Italy, in the section "Statistics", the following notions are studied (Bergamini et al., 2008): population and sample, quantitative and qualitative characteristics, comprehensive research and questionnaires, frequency, frequency table, relative frequency, statistical data, population and sample, qualitative and quantitative characteristics, studies of the total population v/s sample studies, sample survey with a questionnaire, frequency, frequency table, relative frequency, percentage, statistical series, discrete and discontinuous quantitative characteristics, two entries table, graphic presentation of data, histogram (polygon of frequencies), areogram (pie chart), Cartesian diagram (points related to segments), ideograms and cartograms, central position indicators (arithmetic mean, weighted arithmetic mean, median, mode), indicators of variation (variance), average deviation from the mean (as an absolute value), standard deviation, normal distribution, statistical estimation and standard error of estimation, confidence interval with a probability of 99.74% (when assessing an arithmetic mean mean,), standard error when estimating a percentage (p), statistics with Excel.

Slavchova-Bozhkova & Kamova (2012) have explained in detail the content of curricula in statistics taught in schools in England. At age 9 - 10 (primary school in Bulgaria), students in England study average values, statistics, tables and graphs. At age 11-13 (corresponding to 5th-7th grades in Bulgaria), they study the methods of sample surveys, of data collection for sample surveys, probabilities, tree diagrams and the Venn diagram. An important point mentioned by the authors is that the knowledge acquired by students in statistics is applied to other subjects studied in school, such as biology, chemistry and geography.

At age 14-16 (corresponding to 8th - 10th grades in Bulgaria – secondary school), students in England work with statistical databases using computer equipment. At age 16-18 (10th -12th grade in Bulgaria), depending on the choice of their future professional development, students have the opportunity to increase their competence in statistics with knowledge on the different types of distributions, studying dependencies (correlation), or calculation of the probabilities of an event.

In the curriculum in some schools in the United States, statistics are taught in grades 6 through 12 (Newton Public Schools). In the 6th grade, the curriculum includes: distributions, probabilities, variance and scatter plots. The 7th grade includes the basics of sample surveys, the concepts of "population" and "sample", the use of samples for obtaining information about the population, representative samples and the conditions required for a sample to be representative. Students also study how to calculate probabilities of complex events by using lists, tables, tree diagrams and simulations. In the 8th grade, the dependencies between two quantitative quantities are studied - linear dependence, strength of interdependence (correlation), straight and inverse dependence, In these 3 years (6th, 7th and 8th grades) the material in Statistics and probabilities occupies about 1/6 of the total material in mathematics.

In 9-12 grades, the students are encouraged to develop skills that will allow them to critically analyze various types of information, for example, in the field of medical statistics or opinion polls.

The information about the curricula in the other countries presented here is obviously quite short, and not sufficiently detailed or complete. Nevertheless it provided a ground for reflections about some necessary future developments of the respective curricula in Bulgaria.

When comparing the teaching material in probabilities and statistics, included in the mathematics curriculum in Bulgaria with that in the other countries (France, Belgium, Italy, England and the United States) it becomes clear that in Bulgaria this material is less developed, and needs further improvement.

SOLUTIONS AND RECOMENDATIONS

The proposals made in this chapter are based on some ideas of the authors that emerged from their experience in teaching statistics in universities, and on their research of the mathematics curricula in Bulgaria and in some other countries, more precisely in the section of statistics. As a result the following proposals can be formulated for further improvements in this area:

- There is an urgent need for improvement of the mathematics curricula in the section for statistics and probabilities. In every grade the knowledge from the previous grades has to be revised, and the new notions should be introduced gradually, using appropriate examples and field experiments, and with special attention to the students' capabilities and individualities. Upon completion of secondary education, the mastery of certain knowledge must be required.
- The authors suggest that the section "Probabilities and statistics" should be redesigned in a comprehensive and coherent way, setting the goals to be achieved at the end of each year of schooling. In every grade, the teaching material should follow the general logic of gradually covering more and more complex notions and methods, from more intuitive expressions in the first grades, to more scientific definitions in the upper grades. For the younger students, it is crucial to start the explanations with examples from everyday life, stimulating their natural ability and interest for better understanding the respective phenomena (classifying objects in groups by a certain feature; finding the predominant feature in a group of objects; trying to identify a tendency, and so forth.)
- Specialized psychological research is necessary to study the aptitude of the young children in primary school to learn probabilities and statistics. The present research leads to the conclusion that teaching material in probabilities and statistics should be included in the school curricula from the 1st grade and even in kindergarten. In each higher grade, the knowledge should be upgraded and new concepts should be introduced. All this requires specialized pedagogical competences, which should be applied when the respective curricula are implemented.
- Efforts have to be made to increase the knowledge and the capacities of schoolteachers to explain the teaching material in probabilities and statistics. In this direction especially, it is appropriate to organize special courses in order to transmit to them the existing teaching methods, new tools and practices for empowering the students with statistical knowledge and skills to solve statistical problems in a specific practical context.

- The main notions studied in the primary and basic grades (*part of a whole, share, percentage, arithmetic mean*) should be consolidated in secondary school, while the new notions should emerge on the basis of previous knowledge.
- In every grade of secondary school, starting from 8th grade, the teaching material should include at least two parts: descriptive statistics and probabilities (including statistical inference, starting from 10th grade). Alongside the study of the measures of descriptive statistics (applicable to stock data), the students should also be acquainted with the basic measures of dynamics (time-series) and growth (absolute growth, growth rate). The teaching material of time-series should be accompanied by extensive practical examples, demonstrated on graphs. A key notion to be explained in this section is the trend.
- The teaching material in "Probabilities and statistics" for 10th grade about population and sample, as well as the main concepts of descriptive statistics, such as median and mode, should be studied in earlier grades. In the 10th grade, if not earlier, this knowledge should be enlarged by introducing the notions of variance and standard deviation, alongside some other measures of statistical dispersion, such as range, outliers, quartiles, deciles, interquartile range. Some of the properties of normal distribution should be taught in connection with the study of variance and the standard deviation.
- The curriculum in "Probabilities and statistics" should support higher education and benefit all young people by providing knowledge on essential interpretative schemes applicable to everyday life. Graduates who will attend university should be prepared for the requirements of higher education. Students who leave school with only a secondary education should be equipped with enough competence to correctly interpret the information from media, and to operate reasonably well in any social communication, including on their work place, in organizing family life, and in all other activities where they will have to deal with mass phenomena.

The notions taught to all students at any level should be congruent with those used in university curricula where applied statistics is taught to students in social and natural sciences.

In the higher grades, the school curricula should make reference to the databases from official sources - National Statistical Institutes, Eurostat, UN, OECD, World Bank, empirical data, and examples from real life. This may be very useful for attaining the goal of cross-curricular links between the mathematics and other subjects studied in school, such as history, geography, economics, biology, chemistry, physics and astronomy, English, philosophy, and even sports.

- New teaching methods have to be developed and introduced in order to overcome some negative attitudes of students for statistics, and to increase their psychological aptitude to understand this material. In certain instances, in some schools or in some groups of pupils, the teaching material may have to be adapted to the specific needs of the students, and to be transmitted to them by specific teaching methods. Such specific problems will certainly require additional efforts in order to achieve optimal results.

CONCLUSION

The mathematics curricula in the section "Probabilities and statistics" has been undergoing profound changes in recent years in Bulgaria, as new teaching material is being included every year in some of the grades. In this process, some disproportions can be noticed, especially concerning gradually introducing new knowledge (concepts, indicators and methods) every year.

The Ministry of education and science should elaborate a new unified platform in order to help further reform in the mathematics curricula in Bulgaria, in particular that which concerns the teaching material in "Probabilities and statistics". The new platform should be conform to the trends and tendencies in the school education in statistics in the era of Education 4.0. It should be based on a clear concept of how to teach statistical thinking to children from a very young age, in elementary school, and how to further introduce to them statistical concepts in a gradual and coherent manner. Appropriate examples should be provided to them, and they should be stimulated to produce and analyze statistical data from their first years in school. Teachers in mathematics should use modern technologies, and schools should be equipped with the necessary tools for the application of modern education methods. They should provide possibilities for data processing and analyzing, for creating and transmitting graphical representation, including interactive ones; effective online and offline education, allowing easy communication in teaching; the possibility of each student to be active in the learning process, and have easy access to the instructor; effective communication between the students, organizing team work in groups.

The general conclusion of the research is that there is a serious unmet need of knowledge on probabilities and statistics in Bulgaria, largely due to insufficiencies in the basic and high school curricula. In the very last years, substantial changes can be noticed in the mathematics curricula for different grades of gymnasium, especially in the section Probabilities and statistics, but they don't seem to follow a coherent scheme aiming to achieve specific educational goals.

Many of the problems of teaching statistics in the schools in Bulgaria are already subjected to serious discussions among the specialists, and there is evidence that the curricula are changing in a positive direction. However more research is needed to support these steps, and to increase their effectiveness.

The future of education in general, and more specifically education in statistics, is obviously linked to the use of modern smart technologies, e-learning, internet, and other practicalities adapted to individual needs and interests of the users, the educators and of the students. One of the features of Education 4.0 – the education in the era of the Fourth Industrial Revolution, is life-long learning. This is linked to a dramatic increase in life expectancy taking place in the present period, and especially in healthy life expectancy. People living on the average 100 years or more will probably become the norm of the future society. Old age will not be associated so much with illnesses, and health problems, because diseases will be treated more effectively by advanced medical methods and technologies. On the other hand, progress in healthy life expectancy will come as a consequence of a mass transition to healthy lifestyles in all ages, starting from early childhood, through life in education, in work, and after official retirement. In all stages of life, education will take place in one form or another. Active life will last much longer, and the average person will be able to change several professional occupations; therefore education will take place also in older ages, and not just in youth as it is today. All these changes can be facilitated by updating the school teaching and learning process, namely by introducing effective teaching methods and techniques, adapted to prepare the young people for the needs of Industry 4.0. One important contribution in this direction undoubtedly is the reform of the mathematics school curricula, especially in the section of statistics.

ACKNOWLEDGMENT

The authors would like to thank Michael Karman for his invaluable help in editing this paper.

This research received no specific grant from any funding agency in the public, commercial, or not-for-profit sectors.

REFERENCES

Batanero, C., Burrill, G., & Reading, C. (2011). *Teaching statistics in school mathematics—challenges for teaching and teacher education*. Springer. doi:10.1007/978-94-007-1131-0

Batanero, C., & Díaz, C. (2010). Training teachers to teach statistics: what can we learn from research? *Statistique et enseignement, 1*(1), 5-20.

Bergamini, M., Trifone, A., & Brozzi, G. G. (2008). Matematika verde - 1. Zanicelli.

Callingham, R., & Watson, J. (2017). The development of statistical literacy at school. *Statistics Education Research Journal, 16*(1), 181–201. http://iase-web.org/documents/SERJ/SERJ16(1)_Callingham.pdf

Dejean, D. (2018). *Maths. Enseignement obligatoire et de spécialité (ES). Enseignement de spécialité (L)*. Hachette Livre, Vanves Cedex.

Gal, I. (2002). Adults' statistical literacy: Meanings, components, responsibilities. *International Statistical Review, 70*(1), 1–25. doi:10.1111/j.1751-5823.2002.tb00336.x

Garfield, J. (1995). How Students Learn Statistics. *International Statistical Review, 63*(1), 25–34. doi:10.2307/1403775

Garfield, J., & Ahlgren, A. (1988). Difficulties in Learning Basic Concepts in Probability and Statistics: Implications for Research. *Journal for Research in Mathematics Education, 19*(1), 44–63. doi:10.2307/749110

Haure, J.-P., & Lavandier, F. (2018). *Annales Bac, 2018. Sujets et Corrigés*. Hachette Livre, Vanves Cedex.

Ministry of Education and Science of Bulgaria. (2016a). *School Educational Programme in Mathematics for the 2nd grade*. https://www.mon.bg/bg/1997

Ministry of Education and Science of Bulgaria. (2016b). *School Educational Programme in Mathematics for the 6th grade*. https://www.mon.bg/bg/1998

Ministry of Education and Science of Bulgaria. (2016c). *School Educational Programme in Mathematics for the 7th grade*. https://www.mon.bg/bg/1690

Ministry of Education and Science of Bulgaria. (2016d). *Educational Programmes in Mathematics by School grades*. https://www.mon.bg/bg/28

Ministry of Education and Science of Bulgaria. (2017). *School Educational Programme in Mathematics for the 3rd grade*. https://www.mon.bg/bg/1689

Montoya, S. (2018). *Defining Literacy. GAML Fifth Meeting*. UNESCO Institute of Statistics. http://gaml.uis.unesco.org/wp-content/uploads/sites/2/2018/12/4.6.1_07_4.6-defining-literacy.pdf

Moore, D. S. (1997). New Pedagogy and New Content: The Case of Statistics. *International Statistical Review*, *65*(2), 123–165. doi:10.1111/j.1751-5823.1997.tb00390.x

Newton Public Schools. https://www.newton.k12.ma.us/

OECD. (2018). *PISA 2015 Results in Focus*. https://www.oecd.org/pisa/pisa-2015-results-in-focus.pdf

Schield, M. (2011). Statistical literacy: A new mission for data producers. *Statistical Journal of the IAOS*, (27), 173–183. https://www.researchgate.net/publication/228968797

Schleicher, A. (2019). *PISA 2019: Insights and Interpretation*. OECD.

Sharma, P. (2019). Digital Revolution of Education 4.0. International Journal of Engineering and Advanced Technology. *International Journal of Engineering Research & Technology (Ahmedabad)*, *9*(2).

Sugareva M., Mourgova, M. (2017, June). *Учебният материал по вероятности и статистика в учебните програми в средните училища в България* [The teaching material in Probabilities and statistics in the curricula of secondary education in Bulgaria] [Paper presentation]. Conference of the Union of Bulgarian Mathematicians "Pedagogical and Technological Aspects of Educational Computer Games", Plovdiv, Bulgaria

Sugareva, M., & Mourgova, M. (2018, April). Образованието по статистика в училищата в България – предпоставка за статистическа грамотност сред населението [School Education in Statistics in Bulgaria – a prerequisite for statistical literacy of the population]. In *Proceedings of the Forty-seventh Spring Conference of the Union of Bulgarian Mathematicians* (pp. 259 – 265). http://www.math.bas.bg/smb/2018_PK/tom_2018/index.html

Tishkovskaya, S., & Lancaste, G. A. (2012). Statistical Education in the 21st Century: A Review of Challenges, Teaching Innovations and Strategies for Reform. *Journal of Statistics Education: An International Journal on the Teaching and Learning of Statistics*, *20*(2). Advance online publication. doi:10.1080/10691898.2012.11889641

Tsakiridou, H., & Vavyla, E. (2019). Probability Concepts in Primary School. *American Journal of Educational Research*, *3*(4), 535–540. doi:10.12691/education-3-4-21

Watson, J., & Callingham, R. (2003). Statistical literacy: A complex hierarchical construct. *Statistics Education Research Journal*, *2*(2), 3–46.

ADDITIONAL READING

Agresti, A., & Franklin, C. (2009). *Statistics: The Art and Science of Learning from Data* (2nd ed.). Pearson Prentice Hall.

Andrew, L. (2009). Experimental probability in elementary school. *Teaching Statistics*, *31*(2), 34–36. doi:10.1111/j.1467-9639.2009.00362.x

Ashline, G., & Frantz, M. (2009). Proportional reasoning and probability. *Synergy Learning*, (Nov/Dec), 8–10.

Ben-Zvi, D., & Garfield, J. (2004). *The Challenge of Developing Statistical Literacy, Reasoning, and Thinking*. Springer. doi:10.1007/1-4020-2278-6

Fischbein, E., & Schnarch, D. (1997). The evolution with age of probabilistic, intuitively based misconceptions. *Journal for Research in Mathematics Education*, *28*(1), 96–105. doi:10.2307/749665

Gurlen, E. (2015). An Analysis of Mathematics Curriculum at Secondary Level. *Social and Behavioral Sciences*, *174*, 1404–1407. doi:10.1016/j.sbspro.2015.01.767

Jones, G. (2005). *Exploring Probability in School: Challenges for Teaching and Learning. Mathematics Education Library*. Springer. doi:10.1007/b105829

Keser, H., & Semerci, A. (2019). Technology trends, Education 4.0 and beyond. *Contemporary Educational Researches Journal*, *9*(3), 39–39. doi:10.18844/cerj.v9i3.4269

Schield, M. (1998). Statistical Literacy and Evidential Statistics. In *Proceedings of the Section on Statistical Education*. Joint Statistical Meetings, American Statistical Association.

KEY TERMS AND DEFINITIONS

Digital Technologies: Devices, methods, and systems using invention of science and the practical use of digital or computerized methods, including internet.

Probability: The degree of belief or knowledge that an event has occurred or will occur. In terms of numerical expression, the probability is always between 0 and 1. The higher the probability of an event, the more certain we are that the event will happen.

Statistical Curricula: A section of the mathematics curriculum in schools, dealing with the study of probabilities and statistics.

Statistical Data: Numerical information, collected by special statistical procedures, in order to provide a true picture of a piece of reality.

Statistical Literacy: The ability to read, understand, compute and analyze statistical data, including tables and graphs.

Statistical Methods: Standardized procedures of data analysis for accomplishing a substantial result, using statistical data.

Statistics Education: The practice of teaching and learning of statistics within the formal or informal education.

Chapter 6
Proposal Intervention Based on the Classroom Project:
STEAM Project at the Instituto Politécnico Nacional

María Elena Zepeda Hurtado
Instituto Politécnico Nacional, Mexico

Alma Alicia Benítez Pérez
Instituto Politécnico Nacional, Mexico

Betsabé Adalia Contreras Domínguez
Instituto Politécnico Nacional, Mexico

ABSTRACT

The objective of the chapter is the elaboration of an intervention proposal based on the STEAM Integrator Project that from mathematics addresses the issues related to the 2030 Agenda proposed by the United Nations. The central element of the proposal is to consider the context and interest of the student for the selection of a topic related to science to explain it, argue it, and describe it in order to model it from different disciplines of the physical-mathematical area in the Upper Middle Level of the Instituto Politécnico Nacional.

DOI: 10.4018/978-1-7998-3868-5.ch006

INTRODUCTION

For its acronym in English STEAM (Science, Technology, Engineering, Arts and Mathematics) is a global trend that intends to train students in engineering or science since there is a downward trend in graduates in these areas, so which requires strengthening education in order to find a variety of teaching strategies where the principles of mathematics, science and engineering are applied in the classroom in an interdisciplinary way. Its purpose is to achieve the training of students that the Fourth Industrial Revolution requires to take advantage of information technologies, robotics and sensors, mainly.

The STEAM methodology is put into practice in the classroom through active methodologies whose challenges are addressed through Project Based Learning (PBL) among others with the intention of bringing the student closer to the development of thought in a practical, creative and interactive way, complex skills and research, analysis, synthesis, experimentation, creativity, argumentation as self-managed and self-directed learners, who solve or propose the solution to real problems for sustainable development.

In Mexico, the Instituto Politécnico Nacional (IPN) has a history of integrative, interdisciplinary and multidisciplinary projects, which through the implementation of the Aula Project (AP) defined as a strategy whose principles are based on the ABP, meaningful learning and the development of competences through collaborative and autonomous work.

In the AP methodology, students select a topic of their interest and propose alternatives or the solution of practical or simulated problems. Consequently, for the implementation of the PBL at present, an organization of spaces, materials and resources is essential, as well as the inclusion of ICT for an effective impact on the development of the student's competences.

This chapter makes a proposal for the intervention of a didactic and innovative integrative project that from mathematics issues related to the 2030 Agenda proposed by the United Nations (UN) and within the context and interest of the student, chooses a topic related to science, addresses a contextualized problem to explain it, argue it, describe it and model it from different disciplines of the physical-mathematical area, in the Upper Middle Level of the IPN.

BACKGROUND

The imminent change in the industry is evident in its impact on various production scenarios worldwide, a clear example is in the progress of the Industrial Revolutions, patent in the characterization that Del Val (2016) makes of industrial revolutions:

The first Industrial Revolution (18th and 19th centuries) is characterized mainly by the mechanization of agricultural and artisanal production processes. The second Revolution (20th century) is defined by serial production that allowed manufacturing products for mass consumption, consequently, the third Industrial Revolution (late 20th century), a transformation that occurs with the rise of electronics and computing, a characterized by the automation of production lines and machines that perform repetitive tasks. The Fourth Industrial Revolution takes advantage of information technologies, robotics and sensors, resulting in the Internet of Things (IoT).

Changes in the industry, in the forms of production and the economy require trained personnel to adapt, and of course, the training processes of the personnel who will work in the industry require current and appropriate training. In the World Economic Forum on Latin America (2018), the current challenges of the Fourth Industrial Revolution are considered for sustainable development such as the distribution of government services in a transparent way, machine learning to connect job seekers with job offers, the precision medicine to improve healthcare outcomes, or Internet of Things sensors to use water more efficiently. However, the greatest challenge of the Fourth Industrial Revolution is the development of talent and skills necessary for professional retraining for job creation, opportunity generation and economic development.

Within the characterization of Industry 4.0, among other findings Basco, Beliz, Coatz and Garnero (2018, p 18) find that the so-called soft-hard hybrid skills will be necessary, citing a study carried out by the Task Force on Digital Economy of the G202 addressing the demand in different sectors of the industry will be nine cognitive, non-cognitive and social skills: literacy; arithmetic; ICT-related skills; STEM skills (science, technology, engineering and math); marketing and accounting; management and communication; Problem resolution; self-organization; predisposition to learn. It is mentioned that ICT skills, numeracy and quantitative STEM skills, as well as self-organization and management and communication skills seem to be especially recognized and paid in the most digitized sectors. This scenario implies opportunities and challenges associated with the Fourth Industrial Revolution.

Considering that the greatest challenge of the Fourth Industrial Revolution is the development of the talent and skills necessary for professional retraining for job creation, generation of opportunities and economic development, the General Assembly of the UN (2015) adopted the so-called Agenda 2030 with 17 goals to be achieved in 15 years spanning the economic, social and environmental spheres: ending poverty, hunger and food security, health, education, gender equality and empowerment of women, water and sanitation, energy, economic growth, infrastructure, reducing inequalities between and within countries, cities, sustainable production and consumption, climate change, oceans, forests, desertification and biological diversity, peace and justice and partnerships.

One of the strategies to contribute to the formation of human capital that contributes to solving problems is the implementation of STEAM education, which aims to involve and take advantage of the points that the areas of Science, Technology, Engineering, Arts and Sciences have in common. Mathematics, what they intend is under an interdisciplinary approach to integrate it into the teaching-learning process, in various contexts and everyday situations to solve complex problems.

Delgado (2019) establishes that the term STEM (for its acronym in English) is the acronym for the terms in English Science, Technology, Engineering and Mathematics (science, technology, engineering and mathematics), identifies that in recent years that different aspects have begun to emerge: STEAM, for example, where the "A" is for the Arts or ST2REAM where "T2" is for teaching or thematic instruction (in English), "R" for reading and "A" for arts.

According to UNESCO (2020), the steady decline in youth enrollment in science and the shortage of engineers is cause for concern. More young people must choose engineering or science as a career and making that choice depends on access to the necessary science, math, technology and engineering (STEM) curriculum.

Studies by UNESCO (2017) state that to education for girls and women in science, technology, engineering and mathematics, only is a 35% of students enrolled in studies related to STEM in higher education are women, 3% of higher education students, for example, choose to study in the field of information and communication technologies (ICT).

The STEM educational movement seeks to increase scientific-technological vocations and connects different tools, perspectives and didactic methodologies. Project-based learning (PBL) is a privileged methodology for the realization of objectives (Doménech, Lope & Mora, 2019).

The difficulty in teaching - learning mathematics is known, so the importance of the STEAM project is to bring the student closer to mathematics, contextualizing the discipline to the student's environment through the approach and solution of problems of interest to the student. Through constructivism. Given which, Zambrano (2017) mentions that it is essential to promote mechanisms to strengthen the teaching-learning processes that encourage the use of technological tools and scientific methods capable of strengthening knowledge and solving the needs of the environment through learning functional.

In this way, the STEAM project is the ideal setting to leave behind the traditional teaching of mathematics. As mentioned by the National Council of Teachers of Mathematics (NCTM, 2000) cited by Alsina (2020), it is one of the organizations that has most promoted this competence approach that emphasizes the learning of mathematics through processes (thinking and doing, as it is known in the Anglo-Saxon literature), compared to a teaching centered on the memorization of definitions and procedures. Thus, five mathematical processes have been established: problem solving; reasoning and proof; the communication; the representation; which in turn introduce connections typical of mathematics, such as reasoning, arguing, discovering, representing, modeling, demonstrating, connecting.

The study by García and García (2020) on the relationship between the STEAM project and mathematics established as the main conclusion for high school students is to develop interdisciplinary units, without neglecting the needs and the context that is present to enrich and strengthen multiple knowledge taught from the same topic. They also point out that the different activities and projects that are generated through the interdisciplinary of subjects develop the ability to work in a team, oral and written communication, critical thinking, among others; required in current business contexts and in today's society.

For more than two decades, the IPN implemented an educational model focused on learning student whose goal was to educate students to learn to learn. The Educational Model had the challenge of implementing the alternative AP (Materials for Reform, IPN, 2004) to promote a dynamic, innovative, collaborative and interdisciplinary methodology, the main axis being the solution of a contextualized problem, defined in a project.

Is a methodological proposal was designer to the principles of constructivism, including methodologies and strategies for teaching and learning innovative and creative, which establishes the conditions conducive to active learning, contextualized, integrated and aimed at understanding, providing opportunities to reflect on the experience and apply what has been learned from a situation to be faced (Alonso, 2018).

The projects contribute to raising learning in science and in particular in mathematics, pertinent to the development of capacities and skills in today's world, promoting the modeling of contextualized situations (Abrantes, 1994), that is, the development of cognitive abilities to explain, argue, describe and model mathematically. To meet these needs, STEMA emphasizes an interdisciplinary educational approach where academically rigorous concepts are coupled with the real thing (Domínguez, Oliveros, Coronado & Valdez, 2019).

Exploring and discussing contextualized problems through projects is a strategy for the formation of globalized thinking in the student, as it is conceived as the process of investigating relationships around a topic, allowing the development of relationships with new information to forge their relational learning and comprehensive (Alsina, 1998; Aravena, 2001).

Problem solving is in connection with applications and modeling, that is, it describes the interrelation between the real world and modeling in mathematics. Niss (1989) recognizes modeling as the art of using mathematics in real life, for his part Gómez (2003) identifies three fundamental elements in a mathematical model, real world objects, set of mathematical expressions and the relationship between them. Thus, Lesh and Lehrer (2003) refer to the processes that intervene during the mathematical interpretation and description of situations for decision making, which constitutes a focus on relationships, patterns and characteristics, therefore, it must focus on the structural characteristics of this situation.

In this context, cognitive skills are the basis of cognitive operations that occur in the learning activity in mathematics, and contribute to structure the knowledge achieved and reasoning (Aragón, 2007). The cognitive components of problem solving converge in the understanding, representation of the content of the problem, the application of solution strategies and the application of processes of both self-regulation and metacognition in order to monitor the follow-up of the resolution-oriented process of the situation (Funke, 2010).

From the above, it is important to consider the curricula conventional as a scheme of distribution discipline and content by grade level, trying to standardize teaching, but this process says very little about what actually happens in classrooms and disciplinary contents.

Thanks to the fact that the objectives of education have evolved towards interdisciplinary learning, the development of skills, the use of technology among others, which displace the set of rigid knowledge focused on the mastery of techniques and the development of mechanical skills, now it is substituted by the so-called higher order intellectual abilities and the formation of attitudes that favor independence, autonomy and responsible decision-making, in changing situations and uncertainty, such as those currently facing the individual in the personal, citizen and professional.

This is because one of the objectives of the IPN of Higher Secondary Education is to ensure that students develop dynamic learning, allowing them to face situations, both familiar and unpublished, in which they require the production or use of ideas, which give rise to an assessment of these situations, and manage to define various options with their respective costs and benefits.

To organize complex learning from assumptions qualitatively different from those on which our training is based, we need to identify the underlying knowledge, skills, attitudes and values that we must review.

Mathematics is the basis of many of the technological scientific branches, it is for this reason, it is a fundamental pillar in the educational process, playing an important role in solving problems within its social, academic and cultural environment, despite the ideas that one has, apparently mathematics and these problems seem to be unrelated, being able to approximate real situations to mathematics learning units to address problems that allow us to model said situations, thus allowing students to achieve meaningful learning .

The ABP's work methodology is based on the development of competencies that result in learning through activities that allow it to reflectively show its expertise to similar situations within daily life.

The role of the teacher and facilitate learning through an organized planning that will lead to achieving meaningful learning and autonomous students. This didactic approach allows incorporating the method of continuous problematization, formulating conjectures and reviewing systems of the acquired knowledge, supported with audiovisual and technological resources, respecting the object of study between the student and the knowledge. This process must ensure that the student develops skills, knowledge and attitudes, in general, research for problem solving and through processes to communicate their thinking and gradually become accustomed to the various means of mathematical expression.

PROPOSAL DIDACTIC

The intervention proposal of the didactic project, focused on teaching and learning through the practical methodology that will be done in a group of sixth semester, end of the technical career, considering the first programmatic unit that addresses statistics as a central theme, the students will be the ones who choose a topic of their preference to carry out the statistical analysis

The objective is: To develop an intervention proposal using the STEAM and PA methodology from the Probability and statistics learning unit for students who are studying the technological baccalaureate at the IPN through the development of projects developed by integrative, interdisciplinary and multidisciplinary students.

The process starts with the choice of the topic, to continue the approach the problem, delimitation and justification, the questions that make up the survey, application, analysis of results and elaboration of the report that integrates conclusions and representation ones of these conclusions using an infogram or infographic.

In table 1 "Programmatic description", the programmatic contents and competencies corresponding to the Probability and statistics learning unit can be identified.

Table 1. Programmatic description

Characteristics	Description (what, how, for what, with what)
Learning unit	Probability and statistics
General Competition	Solves problems related to descriptive statistics and probability in his academic, social and global environment.
Particular Competition	The student uses descriptive statistics in solving problems that arise in his academic, social and global environment.
RAP (Proposed Learning Outcomes)	1. Prepares tables of frequency distribution and data graphs, obtained from a sample or population, for posing problems in the academic, social and global fields. 2. It calculates the measures of central tendency, position and dispersion, in data obtained from a sample or a population in solving problems in the academic, social and global fields.

Source: IPN (2008) study program Probability and statistics.

The development of the AP presents a defined methodology integrated into stages that range from planning, agreements between students and teachers of the various learning units for the solution of a problem in a practical way, until reaching agreements on learning evidence and evaluation criteria.

Table 2 shows the breakdown of the stages indicated normatively in the AP; the first column exposes characteristics, the second column describes how the learning units are related and the contribution to the project.

From this proposal, the students will verify the use of the data obtained, of interpreting them verbally and orally, as well as of exposing those using different languages: graphics, in Spanish and another language, in a creative way using technology as a resource. To investigate, process information and use various programs to make the graphic diagram, for the elaboration of infogram use free templates that are available on the internet, as well as in the Office package, which is a great help for them.

The limitations that the implementation of the STEAM methodology may have is that when it is considered an innovative methodology, much is unknown about it and although within the framework of the IPN there is a history of the AP, the knowledge and collaborative work of the teachers is still in doubt to integrate their learning units into a common project, which requires training, training and updating in various areas: pedagogical, disciplinary and technological.

Table 2. Description of the AP stages

Proposal Intervention Based on the Classroom Project

Stage	Description	Learning units
1st	Teachers meeting, to plan the project integrating the learning units involved.	Assembly for the choice of theme and delimitation. Units of learning involved: Probability and Statistics, Basic Computing Philosophy, English, Communication and leadership Basic Biology, History, Youth and career guidance, Physics, Chemistry
2nd	Agree to develop skills and learning to achieve during the semester. Each learning unit involved defines the activity, product and evaluation	**Statement of the problem in context** Choose a problem in context: academic, family, national community in relation to the 2030 agenda: to the economic, technological and social environment, causes and consequences
3rd	Definition of the project theme or work hypothesis, adjusts your didactic planning of the academy. The participation of each teacher in the development of the project is defined, generic competences will be developed from their learning unit and what are the disciplinary competences, activities, strategies and / or products, the time in which they are addressed and with which they are going to develop the project.	Topic: economic, ecological, energy, social, cultural and scientific. Participants: 40 to 50, the members of a group of students between 15 and 16 years old 9 teachers (out of 9).
4th	Meetings to link the activities between the learning units, make a schedule to schedule the follow-up sessions, preparation of the preliminary project and delivery of required reports.	Documentary research, focused on the theme of variation for the modeling of a real situation through group projects, involving a real problem. The process of building the mathematical model.
5th	Development of planned activities, monitoring and review of compliance with expected learning, in order to make adjustments.	Documentary investigation of each of the aspects indicated in the learning units. Treatment and conversion of graphic representations from the analysis of the data obtained from its variation behavior, to identify the variables and establish the articulation to build the mathematical model of the situation. Making of agreements and decisions continuously
6th	It consists of integrating the final product, disseminating the results and evaluating it. This evaluation is based on the competency approach, so it integrates different types of learning, it is comprehensive in nature.	Treatment and conversion of graphic representations from the analysis of the data obtained from its variation behavior, to identify the variables and establish the articulation to build the mathematical model of the situation. Preliminary construction of the report, whose structure is justification, study objective, research question, theoretical framework, methodology, analysis and results.
END	Final product	Written report, elaboration of the infographic and oral presentation.

Source: Own elaboration

FUTURE RESEARCH

It will be necessary to carry out future research on various aspects of science teaching and learning, on the one hand those imminently related to students as the origin of the lack of interest and motivation to learn specifically the sciences related to STEAM considering as a basis the mathematics, and on the other, research on teacher training and formal updating and its impact in the classroom, without neglecting the results in the classroom.

Everyone has to convince himself or herself that this in-depth review of their work is worth undertaking, because it involves a very high degree of questioning. However, it can also lead us to adopt a new perspective full of surprising challenges.

Therefore, it is important to carry out research that allows us to recognize the complexity of the problems faced in STEAM training at the upper secondary level. In addition, the implementation of proposals oriented towards the generation of feasible, flexible and durable solutions that consider the real times required for academic achievement in students is relevant. Thus, it is important to recognize that there are no recipes for mechanical application for teaching a subject, but there are the capacities of a teacher who possesses openness, flexibility, commitment and interest to innovate from the classroom in order to promote meaningful permanent learning.

REFERENCES

Abrantes, P. (1994). *O trabalho de projecto e a relaçao dos alunos com a matemática en expreriência do projecto MAT789.* Tesis doctoral no publicada, Universidad de Lisboa, Portugal.

Alarcón, J., Bonilla, E. Nava, R., Rojano, T., & Quintero, R. (1994). *Libro para el maestro. Matemáticas.* México: SEP.

Alonso, A. (2018). Aprendizaje basado en proyectos para el desarrollo de la competencia digital docente en la formación inicial del profesorado. *Revista Latinoamericana de Tecnología Educativa, 17*(1), 9–24.

Alsina, C. (1998). *Neither a microscope nor a telescope, just a mathscope. Proceed.* International Community of Teachers of Mathematical Modelling and Applications (ICTMA).

Aragón, M. (2007). Las ciencias experimentales y la enseñanza bilingüe. *Revista Eureka sobre enseñanza y divulgación de las ciencias*, 152-175.

Aravena, M. (2001). Evaluación de proyectos para un curso de álgebra universitaria. Un estudio basado en la modelización polinómica. Tesis doctorado, Universidad de Barcelona, España.

Basco, A., Beliz, G., Coatz, D., & Garnero, P. (2018). Industria 4.0. Fabricando el futuro, 18-19.

Del Val, J. (2016). Industria 4.0: la transformación digital de la industria. *Conferencia de Directores y Decanos de Ingeniería Informática.* Retrieved from http://coddii.org/wp-content/uploads/2016/10/Informe-CODDII-Industria-4.0.pdf

Delgado, P. (2019). *Educación STEM: ¿qué es y cómo sacarle provecho? Observatorio de innovación educativa: tecnológico de Monterrey.* Retrieved from https://observatorio.tec.mx/edu-news/educacion-stem-que-es-y-como-sacarle-provecho

Domènech, J., Lope, S. & Mora, L. (2019). Qué proyectos STEM diseña y qué dificultades expresa el profesorado de secundaria sobre Aprendizaje Basado en Proyectos. *Revista Eureka sobre Enseñanza y Divulgación de las Ciencias, 16*(2), 35-46.

Foro Económico Mundial sobre América Latina América Latina en un punto de inflexión: construyendo una nueva narrative. (2018). Available from http://www3.weforum.org/docs/LA18_Report_Spanish.pdf

Funke, J. (2010). Complex problema solving: A case for complex cognition? *Cognitive Processing, 11*(2), 133–142. doi:10.100710339-009-0345-0 PMID:19902283

García, M., & García, V. (2020). Metodología STEAM y su uso en Matemáticas para estudiantes de bachillerato en tiempos de pandemia Covid-19. *Revista Científica (Maracaibo), 6*(2), 163–180.

Gómez, J. (2003). La Modelización Matemática. Una herramienta válida en la enseñanza de las Matemáticas Universitarias. *Sumas, 42*, 37–35.

Instituto Politécnico Nacional. (2004). *Materiales para Reforma. Un Nuevo Modelo educativo para el IPN.* IPN.

Instituto Politécnico Nacional. (2008). *Programa de estudios de la Unidad de Aprendizaje de Probabilidad y estadística.* IPN.

IPN. (2006). Geometría Analítica. Libro del Profesor. México: IPN.

Lesh, R., & Lehrer, R. (2003). Models and Modeling Perspective on the Development of Students and Teachers. *Mathematical Thinking and Learning, 5*(2-3), 2–3, 109–129. doi:10.1080/10986065.2003.9679996

Nations United. (2015). *Asamblea General.* Retrieved from https://www.un.org/sustainabledevelopment/es/

Niss, M. (1989). Aim and of applications and modeling in mathematics curricula. In *Applications and modeling in learning and teaching mathematics* (pp. 22–32). Ellis Horwood.

Osuna, Mariela, Ruiz, Amparo, Ortega, Alberto, & Salas. (2019). Retos de ingeniería: Enfoque educativo STEM+A en la revolución industrial 4.0. *Innovación Educativa (México, D.F.)*, *19*(80), 15–32.

UNESCO. (2017*) Coloquio Internacional y Foro sobre Políticas de la UNESCO. Descifrar el código: la educación de las niñas en materia de Ciencias, Tecnología, Ingeniería y Matemáticas.* Retrieved from https://es.unesco.org/coloquio-internacional-y-foro-politicas-unesco-descifrar-codigo-educacion-ninas-materia-ciencias

UNESCO. (2020) *STEM Resources.* Retrieved from http://www.unesco.org/new/en/natural-sciences/science-technology/engineering/engineering-education/stem-resources/

Zambrano, K. (2017). Fortalecimiento de las matemáticas a través de las STEAM en la Tecnoacademia de Neiva. *Revista Ciencias Humanas*, *14*, 39–52.

Chapter 7
Teaching and Learning of Mathematics in Lower Primary in Schools in Ghana

Frank Quansah
University of Cape Coast, Ghana

ABSTRACT

Mathematics is a fundamental human activity – a way of making sense of the world. Children possess a natural curiosity and interest in mathematics and come to school with an understanding of mathematical concepts and problem-solving strategies that they have discovered through explorations of the world around them. The purpose of this study is to find out how mathematics is being taught and learnt among primary one pupils in basic schools in Ghana. A qualitative approach was employed to provide answers to the research questions raised. It was found that teachers employed variety of procedures for teaching mathematics. However, there were some problems encountered in the teaching and learning of mathematics.

INTRODUCTION

Mathematics is a fundamental human activity – a way of making sense of the world. Children possess a natural curiosity and interest in mathematics and come to school with an understanding of mathematical concepts and problem-solving strategies that they have discovered through explorations of the world around them (Ginsburg, 2002). Mathematics has historically been a subject that many students struggle with and increasingly dislike as they progress through the grades (Stodolsky, 2008). Teachers should always try to find ways to actively engage their students not only

DOI: 10.4018/978-1-7998-3868-5.ch007

for understanding concepts but also to create elements of fun and excitement so that students' interest can be kindled. Using manipulative materials has become one way of involving students in fun learning that encourages motivation of students. Manipulatives have also been useful in making abstract ideas concrete for learners and thereby making for conceptual understanding.

Quality has been the main aim of many educational reforms, acts and policies in Ghana to improve the academic performance of pupils. It is generally accepted that mathematics education is an important dimension of general education. It is perceived as a core area of learning is probably most education systems throughout the world. However, it is important to know how well mathematics is being taught and learnt in primary one. The ability to develop and apply mathematical thinking to solve a range of problems in the everyday situation is important for all learners. According to the World Bank study in 2011, every extra year of primary education increases a person's productivity (in this case hourly wage rate) by 10 to 30 percent. In addition to being a right, basic education underpins the success of a society. Every year of primary education increases a person's productivity and reduces their dependence on social resources. The goal of education is to enable children to learn and realize their full potential.

Since September 1987, the government of Ghana has embarked upon a new educational programme geared strategically at making education more accessible to all children of school-going age improving equity and the quality of education as a whole and making education more relevant to the socio-economic needs of the country. Enhancing the quality in education, therefore, must be based on developing educational systems that are integrated and responsive to the multiple obstacles to children learning. The benefit and impacts of quality education also make invaluable contributions to all areas of human development.

Also, the syllabus is designed to help pupils to develop basic ideas of quantity and space, use basic mathematics and necessary strategies for solving problems encountered in daily life by recognizing relationships between numbers, and many others. However, the syllabus is structured to cover the first six years of primary school education. Each year's work has been divided into units. The unit topic for each year has been arranged in the sequence in which teachers are expected to teach them. Moreover, the syllabus developers (G.E.S) wish to discourage teachers from forcing the instructional pace but would rather advise teachers to ensure that pupils progressively acquire a good understanding and application of the materials specified for each year's classwork. Also, it is hoped that no topics will be glossed over for lack of time because it is not desirable to create gaps in pupils' knowledge. These topics designed would help pupils to socialize, perform number operations, and make use of appropriate strategies of calculations, adjust to and handle number words and many others.

According to Ghana Education Service(G.E.S) syllabus (2012) some of the topics that are to be taught by teachers in primary one are; pre-number work, groups of objects, counting objects, numbers and numerals 1, addition (up to 5), solid shapes, and many others. Also, mathematics is to be allocated eight periods within a week, each period consisting of thirty minutes. The eight periods should be divided into four double periods each one-hour duration for the week. However, teachers must also ensure that there is maximum pupil participation in the lessons. The aim of the subject can be achieved when teachers create learning situations and provide guided opportunities for pupils to acquire as much knowledge and understanding of mathematics as possible through their activities. Teachers must also try to show, explain and demonstrate to pupils to aid easy understanding. Teacher learner activities will also ensure maximum participation by pupils during lessons. The major part of a pupil's learning experience should consist of opportunities to explore various mathematical situations in their environment to enable them to make their observations and discoveries and record them. Teachers should also help pupils learn to compare, classify, analyses, look for patterns, spot relationships and come to their conclusions. Route learning and drill-oriented methods must be avoided by teachers but rather emphases participatory teaching and learning in his lesson. At the end of lessons, the teacher should ask questions and set task and assignments that will challenge pupils to apply their knowledge to issues and problems.

As the learning and teaching of mathematics are concerned, there are situations that will hinder the progress of the teaching and learning process. The way mathematics is being taught and learnt is of great concern as it can affect the pupils either positively or negatively due to the way it is being taught. It can negatively affect the performance of the pupil if not taught well. However, it is at the primary level that pupils are introduced to the basic concepts in mathematics. Therefore inability to conform to the laydown rules and regulations can affect the pupils involved.

Presently, it is estimated that between four and seven percent of the school-age population experience some form of mathematics-focused disability (Gross-Tsur, Monar, & Shalev, 2006). Bryant, Hartman and Kim (2003) reported that children with difficulties in math often have other related difficulties, such as in-memory, poor calculation skills, number reversals, and difficulty understanding conceptual and/ or procedural processes, especially as represented through symbols and signs. My observations and interaction with the headteachers show that the process by which mathematics is being taught in the early stage of lower primary is inappropriate. This study, therefore, is to examine the methodology in teaching and learning of mathematics. It is hoped that when teachers use correct or appropriate methodology, techniques and skills, pupils' performance in mathematics will improve. Given that, it is pertinent to conduct this study to investigate the problems primary one teacher's face in teaching mathematics, and also, the problems pupils in primary one face in learning mathematics. The purpose of this study is to find out how mathematics is being taught and learnt among Primary one pupils in basic schools in Ghana. Specifically, the study examined the; (1) methods of teaching mathematics in primary one; (2) problems teachers face in teaching mathematics; and (3) problems children face in learning mathematics.

This research is aimed to know whether the right methods are being used or not. The findings of this research are going to influence the teaching and learning of mathematics in Basic School. Also, this research will make us aware of the challenges or problems teachers face when teaching mathematics. Furthermore, the study will serve as a source of reference for other professionals in the field of teaching and learning mathematics at the primary one.

Research Questions

Concerning the study, the following questions were developed to guide the study;

1. What teaching process/methods do teachers employ in teaching mathematics in primary one?
2. What problems do teachers face in teaching mathematics in the primary one?
3. What problems do children face in learning mathematics in the primary one?

Methods of Teaching Mathematics

Teacher's approach to mathematics education is considered as one important factor for children's opportunities to develop mathematical knowledge and skills in their early years (Bjorklund, 2015). In education, one important aspect is how teachers understand a certain subject to be taught and how it is learnt by the children. How one understands a subject reflects how guidelines and curriculum goals are interpreted, what content he/she chooses to work with, and how different methods are used. Such issues are often discussed in terms of pedagogical content knowledge (Ball, Thames & Phelps, 2008), which partly includes knowledge of the subject matter, but even more how the teacher understands the learning of the subject matter and what difficulties may occur, which affect how the teacher organizes the learning activities. Ball et al. (2008) discuss these issues in terms of knowledge of content and students, knowledge of content and teaching, and knowledge of content and curriculum. This means that the teaching practice is a complex act, including the subject conceptualized in pedagogical frames, how the subject is made possible for the children to learn and related to the curricular framework, both national guidelines and preparations made by the teacher.

Woods and Schweingruber (2009) contend that mathematical knowledge is not limited to procedures or factual knowledge; instead, according to a great number of research and theoretical standpoints. Mathematics education in the early years should focus on conceptual knowledge, reasoning skills, communication and mathematical literacy to facilitate more advanced learning in later school years. Cross, Woods and Schweingruber (2009) argue for the necessity to perceive mathematics as both a content of knowledge and content of skills and operations, which expands the learning objective to not only learning about mathematical notions and procedures but also learning to interpret and use mathematics effectively in various situations, for instance, mathematical literacy.

According to Martin, Mullis, Gregory, Hoyle, and Shen (2000) opportunity to learn mathematics effectively is dependent upon a wide range of factors, but among the most important are those which are related to activities and practices within the classroom. The above studies focused strongly on the classroom and what happens there. They suggested that within the classroom, it is possible to discern several key elements. These include teaching practices; the nature of student learning activities; the amount and nature of engaged learning time experienced by students; the learning environment and the scope and nature of the feedback given to students.

Peterson (2008) also listed effective teaching practices as (a) a focus on the meaning and understanding mathematics and the learning task; (b) encouragement of student autonomy, independence, self-direction and persistence in learning; and (c) teaching of higher-level cognitive processes and strategies. This indicates that teaching practices are central to understanding what makes effective teaching. Brophy and Good (cited in Kakai, 2011) identified successful teaching strategies as requiring an organized approach to teaching where the material was taught until it was mastered. They also argued that in the classroom instruction, three modes exist: (i) giving information; (ii) soliciting information; and (iii) providing feedback. This implies that both the teacher and students should participate in the teaching and learning process.

In addition to good sequencing, of information by the teacher, student actions within the classroom have also been identified as key determinants of effective teaching (Raymond & Leinenbach, 2000). Therefore, plans for effective teaching of mathematics to support students' conceptual understanding need to emphasize the need for mathematically challenging and significant tasks (Fraivilling, 2001). This view is also supported by Nickerson (2008) and Stigler & Perry (2008) noted that students' involvement in classroom activities is a major factor affecting student achievement. Mathematics learning mainly constitutes students' engagement in classroom activities.

Some scholars (Peressini, Borko, Romagnano, Knuth & Willis, 2004) suggested that teachers have to engage students in rich, meaningful tasks when teaching. This means students' thinking whether shared orally or in writing, must be used by teachers to guide the classroom in the exploration of important mathematical ideas. Mapolelo (2008) indicated that factors such as the decisions that teachers make, the methods they use, and the attitudes displayed are relevant to performance on mathematical tasks and influences the direction and outcome of student performance. Gitonga (2000) points out that the success or failure of any innovation in teaching ultimately lies on the receptiveness and flexibility of the classroom teacher. This suggests that mathematics teachers have to be conversant with various teaching methods that are appropriate for any given level of learners. Effective teaching requires knowing and understanding learners and pedagogical strategies since students learn mathematics through experiencing what teachers provide. Thus, students' understanding of mathematics, their ability to use it to solve problems, and their confidence in, and disposition towards mathematics are shaped by the teaching they encounter in school (National Council of Teachers of Mathematics in Kenya (NCTM), 2000).

Mackay (1981) suggested that teaching should be learner-centred so that learning becomes more relevant to the learner. Mackay further recommends that schools should teach basic computational skills for problem-solving. This emphasizes the need for the learners' active participation in the classroom activities. This also implies that the teachers' role is to offer opportunities that will lead to student-centred activities in class, which could enhance performance in mathematics. Such opportunities could be realized if teachers used teaching methods that encourage students' participation in class.

Miheso (2002) posited that teaching mathematics well involves creating, enriching, maintaining and adapting instruction to move towards mathematical goals while capturing and sustaining, as well as, engaging students in building mathematical understanding. She further notes that teachers have a wide range of instructional strategies at their discretion, which, often differ in terms of the amount of teacher direction, the number and type of students being taught at any given time. This implies that the choice of any of these strategies depends on their purpose in teaching, their understanding of teaching-learning processes and considerations of cost-effectiveness. Mathematics teachers, therefore, need to be more cautious when choosing teaching strategies to ensure a clear and harmonious link between syllabus contents.

According to a report by the European Commission (2011) on Common Challenges and National Policies indicated that a range of factors influence the way mathematics is taught and learned. Results from the survey suggest that education outcomes are related not only to students' family background but also to the quality of teaching and certain structural and organisation features of education systems. Observations of pre-primary teachers' mathematics teaching give evidence that the character of the practice contributes to children's interest in learning mathematics (Lerkkanen, Kiuru, Pakarinen, Viljaranta, Poikkeus, Rasku-Puttonen, Siekkinen, & Nurmi, 2012).

They further recommended that a child-centred approach, in which mathematics is taught thematically and with respect for children's perspectives, seems to favour children's developing interest in mathematics and learning.

Problems Teachers Face in Teaching Mathematics

Research shows that the lack of confidence of teachers in knowledge content is a major problem devastating the achievement of educational objectives. A study of Swedish preschool teachers by Doverborg, Pramling and Samuelsson (2009) reveals that many early childhood teachers sense that they lack sufficient knowledge of mathematics, with the consequence that they do not feel confident to work with mathematics as a learning object. A teacher who lacks confidence in teaching in a subject area is not able to impart efficiently whatever knowledge he/she is supposed to impart.

Another problem facing most teachers of mathematics is their ability to use the appropriate instructional media to provide students with a situation near to reality. The abstract mathematical concepts are made concrete when students see and do some activities using real objects like models. Muvango (2013) reported that studies indicate that people learn 25 to 30 per cent more when visual aids are used in teaching because they hold attention, motivate to take any action, increase the permanence of learning, make the job of teaching easier and create interest. The availability and utilization of instructional resources for the teaching of any subject in the school curriculum, the mathematics included, is very crucial. However, in a situation where schools are not able to provide all the materials needed for instruction, teachers have to develop teaching materials from the immediate environment through improvisation.

Mathematics teachers have to use more practical strategies and concrete ways to help students grasp mathematical concepts, improve their proficiency and generalize knowledge in multiple contexts (Merlot Pedagogy, 2013). Practical work is important in teaching and learning mathematics. Proper understanding of concepts being taught requires that they be concretized through the use of suitable experimental/practical work, teaching aids and real-life experiences. It is a fact that we are more likely to remember/internalize what we do than what we see or hear. Practical work in mathematics ensures learners' participation and variation of stimuli. Apart from making theoretical and abstract concepts real and concrete, it enhances learning by promoting curiosity and interest in addition to awakening manipulative skills. Abimbade (2007) found out that the use of instructional resources in teaching and learning makes students learn more and retain better what they have been taught and that it also promotes and sustains students' interest. It also allows learners to discover themselves and their abilities.

Schramn (2007) notes that instructional materials enrich learners' knowledge and reinforce verbal instruction. This is because learning is more effective through seeing and touching than through hearing only. Oyeniran (2013) argues that students learn best if they are allowed to see and make observations of what they are taught. He observes that a good instructional material serves as a substitute for real-life objects in the classroom as against the use of the exploratory method.

Philipp (2010) identified two major problems teachers face in teaching mathematics. First, managing a classroom of students (disciplining students, assigning and collecting homework, assessing and evaluating students); and secondly, addressing the needs of individual students. The teaching of mathematics requires special skills from teachers and keen attention from students. Most teachers have been raising issues of a large number of students they handle. The teaching and learning of mathematics require more practice and exercises. Due to the large of students teachers handle, they find it difficult to give more exercises and homework to students, since it is time-consuming. The benefits of exercises and homework cannot be undermined. It is through these exercises that teachers can identify the problems that students are facing.

Problems Children Face in Learning Mathematics

National Center for Education Statistics (NCES) (2004) reported that mathematics is often challenging for students with and without disabilities to master. Children with difficulties in math often have other related difficulties, such as in-memory, poor calculation skills, number reversals, and difficulty understanding conceptual and/or procedural processes, especially as represented through symbols and signs (Bryant, Hartman, & Kim, 2003).

The report of the Expert Panel on Mathematics in Ontario (2004), identified a couple of problems that lower-level children face in learning mathematics. Among the problems, but not limited to are spatial perception, reversals, figure-ground perception, visual discrimination, auditory processing, motor ability, and receptive ability. About the spatial perceptions, children find it difficult to understand the concept of fractions, writing fractions, discerning differences in size and shape. Miller and Mercer (2017) indicated that most children struggle with dealing with directional aspects of mathematics, for example, solving problems involving single addition – up-down, regrouping – left-right, aligning numbers or using the number line.

Further, it has been noted that most children in their early childhood have difficulty maintaining the sense of a problem and not mixing up parts of different problems, reading multi-digit numbers, and copying symbols properly. Concerning visual discrimination, most children have problems with identifying symbols gaining information from pictures, charts, or graphs; being able to use visually presented material productively; telling the difference between a quarter and a nickel; the numbers 6 and 9; the small hand on a clock and the large one; and using many mathematics skills (e.g., in measurement, and estimation). In terms of auditory processing, some problems have been identified with most children. Amongst them are hearing a pattern in counting, deciphering numbers that are spoken (e.g., 30 and 13), performing oral drills, and identifying ordinal numbers. Also, most children have problems with presenting legible work, they have difficulties in connecting vocabulary words with an understanding of mathematical concepts (e.g., first, greater than, times, difference).

Concrete Representational Abstract (CRA) Model

This instructional approach to differentiating instruction in math involves a sequence of instruction (Maccini, Mulcahy, & Wilson, 2007) to master complex concepts in mathematics. CRA involves utilizing manipulatives (concrete). Once the student has mastered the math concept using manipulatives, the objects are replaced with pictorial representations, such as a picture of the object (representational). This level is a critical bridge between the concrete manipulatives and the abstract symbols (e.g., additions, minus, etc.), as this step builds the mental schema bridging these two levels. It is critical to developing mathematics conceptual knowledge during the representational level of learning (Little, 2007). Once the student can comprehend representational figures and designs, Arabic symbols and explanation of the algorithm (abstract) are taught. Successful performance at the abstract level is the goal, as mathematics is most often expressed and assessed at this level. The levels of CRA are sequentially interrelated and interconnected. Each level prepares the student for the next level of learning (Witzel, 2005). The connections between the levels of learning are critical to foster student learning. For example, lessons and units must be designed to be easily represented pictorially and described abstractly. When lessons and units are planned to address the levels of learning, the teacher will be able to differentiate instruction to meet the students' needs for content master.

Constructivist Model

This study was guided by the learning theory based on the constructivist model. This learning theory indicates that mathematical understanding in young children is closely associated with sensory perception and concrete experience. Children begin to understand symbols and abstract concepts only after experiencing the ideas on a concrete level (Piaget, 1952). Manipulatives are effective tools in mathematics education by helping children move from a concrete to an abstract level of understanding. Students who see, touch, take part and manipulate physical objects begin to develop clearer mental images and can represent abstract ideas more completely than those whose concrete experiences are limited (Heddens, 1986). Children whose mathematical learning is firmly grounded in manipulative experiences will be more likely to bridge the gap between the world in which they live and the abstract world of mathematics (Dienes, 2015).

Also, constructivism advances the idea that the individual begins to develop understanding through personal experiences and personal connections. "The interlacing of content, context and understanding, the individual negotiation of meaning, and the construction of knowledge" are promoted in a learning environment that promotes constructivism (Land & Hannafin, 2006). Furthermore, collaboration, real or virtual, which brings about new ways of conceiving concepts that might not be visualized by individual alone are considered integral (Abrami, 2001).

Zone of Proximal Development

Worldwide, performance in Mathematics by students has persistently been poor (Mbugua, Kibet, Muthaaa & Nkonke, 2012). Research suggests that the application of Vygotsky's theory of the Zone of Proximal Development (ZPD) could improve mathematical achievement (Roosevelt, 2008). It is argued that the theoretical significance of the ZPD is that it enables penetration into the causal-dynamic and genetic connections determining the process of mental development (Obukhova, & Korepanova, 2009). This writes up the relevance and implications of the ZPD theory to the teaching of Mathematics.

The zone of proximal or potential development perhaps is the best-known concept of Vygotskian sociocultural psychology. Initially, it was elaborated for psychological testing in schools (Vygotsky, 1962). Vygotsky stated that testing should be based not only on the current level of a child's achievements but also (and mainly) on the child's potential development. The actual level of development (level of independent performance) does not sufficiently describe development. Rather, it indicates what is already developed or achieved, it is a "yesterday of development". The level of assisted performance indicates what a person can achieve shortly, what is developing (potential level, "tomorrow of development", what a person "can be") (Vygotsky, 1978). Thus, the zone of proximal development is the distance between what a person can do with and without help. The term proximal (nearby) indicates that the assistance provided goes just slightly beyond the learners' current competence complementing and building on their existing abilities (Cole & Cole, 2001).

Vygotsky conceives of the zone of proximal development as central to the instructional enhancement and classroom change in Mathematics. According to Vygotsky, the zone of proximal development is the distance between the actual development level as determined by independent problem solving and the level of potential development as determined through problem-solving under adult guidance or collaboration of more capable peers (Murray & Arroyo, 2002). Vygotsky (1978) opined that the ZPD is the current or actual level of development of the learner and the next level attainable through the use of mediating semiotic and environmental tools and capable adult or peer facilitation. In other words, it is what a child can do alone at a particular point in time. The second, potential development" is defined as that which a child can achieve if given the benefit of support during the task. It is the ability to solve problems "under adult guidance or in collaboration with more capable peers.

Inferring or relating ZPD in the teaching of mathematics, the teaching of mathematics goes through three levels. First, the teacher should know what the child can do independently. Secondly, what the child can do with the help of a teacher or an expert (ZPD), and finally, what the child cannot do even with the assistance of an expert. The Zone of Proximal Development (ZPD) indicates the functions that have not yet matured but are in the process of maturation. The ZPD is critical for Mathematics learning and instruction because it determines the level of work to be taught which is desirable to the child. Vygotsky suggested that teaching geared to developmental levels that have already been achieved will be ineffective and that the only 'good learning' is that in advance of development (Wells, 2000).

METHODS

The descriptive survey design was employed using the qualitative approach for the study. The descriptive survey research design involves gathering data to test hypotheses or finding answers to research question concerning the current status of the subject matter (Gay, 2005). The descriptive design was used because the study intends to describe the problems both teachers and learners face in the teaching and learning of mathematics.

Sampling Procedures

The convenient sampling method was used to engage 4 primary one teachers. This number was chosen because the researchers intend using a qualitative approach. In using a qualitative approach, it is advisable not to use large sample size. Again, since the teachers are ones who are in the classroom with these children, they are in a better position to know the problems their students face in learning mathematics.

Data Collection Instrument

The research instrument is the medium through which the researcher gathers information needed in a particular study. There are several modes of instruments for data collection. The choice of a particular instrument is dependent on the research design used, the research questions the study intends to answer, and the sample size involve (Miles, 2001). This study resorted to the use of an interview guide. The interview guide was self-designed based on the research questions.

Data Analysis Procedure

The data collected was analysed qualitatively. The responses provided by the teachers were transcribed and analysed in themes. The results were presented according to research questions. In the thematic analysis, major issues raised during the interview would be put in main themes to make the analysis of the data relatively easier for discussion. Braun and Clarke (2006) explained thematic analysis as a qualitative analytic method for identifying, analysing and reporting patterns (themes) within data. They argued further that as the thematic analysis does not require the detailed theoretical and technological knowledge of approaches such as grounded theory, but offers a more accessible form of analysis, particularly for those early in a qualitative research career. These formed the basis for using thematic analysis.

RESULTS

Background Information of Teachers

It was revealed that out of the four teachers, 2 of them were females, while 2 were males. The youngest teacher was 26 years old whereas the oldest was 54 years old. Two of the teachers had 4 years teaching experience in primary one, 1 had about 10 years' experience, and the other had more than 10 years' experience.

What teaching process/methods do teachers employ in teaching mathematics in primary one?

Teachers were asked to indicate and explain the methods they use in teaching mathematics. Their responses are illustrated below;

A teacher with 4 years' experience said:

Primary one pupils are kids who are very young and have not developed mentally, so in teaching them mathematics, I use the demonstration method. By this I mean, on very topic, I come in with teaching and learning materials to demonstrate to the children. At their age, that is the only way they can learn mathematics. For, instance, in teaching addition and subtraction, I use toffees to do the demonstrations. I realize that these children like toffees a lot, and upon seeing them, they tend to give all their attention... after I have done the demonstration with them, I give room each pupil to also perform it.

Another teacher responded:

...the learner activity method is best for teaching class one pupils. This involves only guiding pupils but allowing them to do the manipulations by themselves. What I have realized is that the children end up sleeping when all the classroom activities are carried out by the teacher. Engaging pupils make them active in class.

Also, one respondent with above 10 years' experience indicated that:

Mostly, I use Ghanaian Language to teach. The children in primary one understand better when they are taught using the local dialect. Also, I use concrete objects in teaching. These children learning better by observing and also manipulating objects by themselves.

From the responses, it is clear that teachers use practical and concrete materials in teaching. They also indicated that teaching should be done in a manner that pupils get involved. These findings are in line with Peressini, Borko, Romagnano, Knuth and Willis (2004), who suggested teachers have to engage students in rich, meaningful tasks when teaching. Again, Mackay (2001) also suggested that teaching should be learner-centred so that learning becomes more relevant to the learner. Mackay further recommends that schools should teach basic computational skills for problem-solving. This emphasizes the need for the learners' active participation in classroom activities. This also implies that the teachers' role is to offer opportunities that will lead to student-centred activities in class, which could enhance performance in mathematics.

What Problems Do Teachers Face in Teaching Mathematics?

The teachers were asked to indicate the problems they face. Their responses are illustrated below;

A respondent with 7 years teaching experience stated that:

A major problem I'm facing is the availability of materials and other resources to enhance teaching. Class one is the foundation level and for that matter pupils in class one must be given maximum attention. The teaching of these pupils should be made practical, but here is the case where there no resources to teach the pupils. This makes teaching very difficult, because you will spend a great deal of energy and time to talk, but these pupils would not understand. This is because everything you are saying is in abstract terms.

Another respondent with over 10 years' experience said:

The class size is too enormous. I'm handling a class of about 45 pupils, and these are children who need special attention. I find it very challenging to mark the exercises of these children. Once the exercises have been marked, these pupils should be guided to effect the corrections of items they had wrong. Due to nature of the class, it makes it impossible. Also, these pupils are also to be taught using the learner-centred activity method. Dues their numbers, only a few of them are able to perform the tasks, and this does not enhance effective classroom learning.

A teacher with 4 years' experience indicated that:

One thing I've observed is that about half of my students have problems with writing the figure '9', '7', and '2'. Repeatedly, they interchange '9' with either 'p', 'b' or '6'. Also, they interchange '7' with '4', or 'F', and finally, figure '2' with '5', and vice versa. I've also noticed that when these pupils are given homework, some of them come back to school with the work undone. This implies most parents do not monitor the activities of their wards, and this affects the teaching of mathematics since old things have to be repeated continuously.

Judging from the excerpts, it can be inferred that teachers are faced with the problem of getting teaching materials and resources, they have a challenge with the large classes they are handling, and also some pupils' inability to write some figures correctly. These findings agree with Schramn (1977) that instructional materials enrich learners' knowledge and reinforce verbal instruction. This is because learning is more effective through seeing and touching materials than through hearing only. Oyeniran (2003) also asserted that students learn best if they are allowed to see and make observations of what they are taught. He observes that a good instructional material serves as a substitute for real-life objects in the classroom as against the use of the exploratory method. Further Philipp (2010) identified two major problems teachers face in teaching mathematics. First, managing a classroom of students (disciplining students, assigning and collecting homework, assessing and evaluating students); and secondly, addressing the needs of individual students. The teaching of mathematics requires special skills from teachers and keen attention from students.

What Problems Do Children Face in Learning Mathematics?

The research question intends to find out the problems pupils face in learning mathematics. The responses of their teachers are illustrated below;

A respondent with 9 years' experience said that:

Pupils have the problem of identifying shapes correctly. Any shape in the form of a square is a box and any circular shape is a circle. Some have problems differentiating the mentioning of figures like 14 and 40.

Another teacher indicated that:

Some people have the people have the problem of deciphering sizes. They cannot differentiate between greater than and less than. They see taller and longer objects as bigger than shorter but wider objects.

It was also reported that:

Most children have problems with writing figures. They put together almost everything they are writing. Some have difficulties in writing figures like 9, 4, 7, 6, 5, and 2, they normally interchange these figures with one another.

From the above findings, it is evident that primary one pupils face a couple of challenges with the learning of mathematics. These findings concur with a report by The Expert Panel on Mathematics in Ontario (2004), which identified a couple of problems that lower-level children face in learning mathematics. Among the problems are spatial perception, reversals, figure-ground perception, visual discrimination, auditory processing, motor ability, and receptive ability. Also, Bryant, Hartman, & Kim (2003) postulated that children with difficulties in math often have other related difficulties, such as in-memory, poor calculation skills, number reversals, and difficulty understanding conceptual and/or procedural processes, especially as represented through symbols and signs.

CONCLUSION

Based on the findings of the study, it can be concluded that primary one teachers are faced with the problem of handling a large number of students. Given that, teachers find it difficult to attend to the needs of all the pupils. Also, it can be concluded that teachers are not able efficiently teachers because of the unavailability of some learning materials. Pupils in primary one are faced with a couple of challenges in learning. We can also conclude that most individuals have problems mathematics because of the challenges they faced during the early years of schooling.

RECOMMENDATIONS

From the findings and conclusions of the study, the following recommendations are made:

1. It is recommended that primary one teachers teach mathematics using concrete objects since that enhances learning.
2. It is recommended that headteachers of schools ensure that classes with a large number of pupils be divided to ease the pressure on the teachers.
3. The education district in Anwiafutu is recommended to ensure that all schools have the necessary instructional materials and resources to enhance teaching and learning.
4. Teachers also to ensure that they give all the necessary assistance pupils they identify as having problems with mathematics.

SUGGESTION FOR FUTURE RESEARCH

It is suggested that future research use longitudinal approach so that the situation will be studied for over a long period. Also, document analysis can be conducted to properly examine the portfolios of pupils to establish the challenges they have in mathematics.

REFERENCES

Abimbade, A. (2007). *Principles and practice of Educational Technology*. International Publishers Limited.

Abrami, P. C. (2001). Understanding and promoting complex learning using technology. *Educational Research and Evaluation*, *7*(2 & 3), 113–136. doi:10.1076/edre.7.2.113.3864

Akubia, L. (2011). *The effect of motivation on the performance of teachers in the Ghana Education Service: A case study of Obuasi Municipality*. KNUST.

Asare, J. K. (2010). *Factors Affecting Recruitment and Retention of Basic School Teachers in Obuasi Municipality*. Academic Press.

Aunola, K., Leskinen, E., & Nurmi, J. E. (2006). Developmental dynamics between mathematical performance, task motivation, and teachers' goals during the transition to primary school. *The British Journal of Educational Psychology*, *76*(1), 21–40. doi:10.1348/000709905X51608 PMID:16573977

Ball, D. L., Thames, M. H., & Phelps, G. (2008). Content knowledge for teaching. *Journal of Teacher Education*, *59*(5), 389–407. doi:10.1177/0022487108324554

Björklund, C. (2015). Pre-primary school teachers' approaches to mathematics education in Finland. *Journal of Early Childhood Education Research*, *4*(2), 69–92.

Braun, V., & Clarke, V. (2006). Using thematic analysis in psychology. *Qualitative Research in Psychology*, *3*(2), 77–101. doi:10.1191/1478088706qp063oa

Bryant, D. P., Hartman, P., & Kim, S. A. (2003). Using explicit and systematic instruction to teach division skills to students with learning disabilities. *Exceptionality*, *11*(3), 151–164. doi:10.1207/S15327035EX1103_03

Cole, M., & Cole, S. (2001). *The development of children* (4th ed.). Scientific American Books. Distributed by W.N. Freeman and Company.

Creswell, J. W. (2012). *Educational research: Planning, conducting and evaluating quantitative and qualitative research* (4th ed.). Pearson Publication.

Cross, C., Woods, T., & Schweingruber, H. (2009). *Mathematics Learning in Early Childhood. Committee on Early Childhood Mathematics*. The National Academic Press.

Dienes, Z. P. (2015). *Building up mathematics*. Hutchinson Education Ltd.

Doverborg, E., & Pramling Samuelsson, I. (2009). Grundläggande matematik. In S. Sheri-dan, I. Pramling Samuelsson, & E. Johansson (Eds.), *Barns tidiga lärande. En tvärsnittsstudie om förskolan som miljö för barns lärande* [Children's Early Learning. A Cross-sectional Study of Preschool as an Environment for Children's Learning] (pp. 125–150). Acta Universitatis Gothoburgensis.

Gay, L. R. (2005). *Educational Research. Competencies for Analysis and Application* (5th ed.). Merrill Publication Company.

Ginsburg, H. (2002). Little children, big mathematics: Learning and teaching in the pre-school. In *Proceedings of the 26th Conference of the International Group for the Psychology of Mathematics Education* (Vol. 1, pp. 3–14). University of East Anglia.

Gitonga, J. M. (2000). *A study of the problems that teacher trainers in primary teachers colleges in the Eastern Province of Kenya face in preparing Trainees to teach Mathematics* (Published M.Ed. thesis). University of Nairobi Press.

Gross-Tsur, V., Manor, O., & Shalev, R. S. (2006). Developmental dyscalculia: Prevalence and demographic features. *Developmental Medicine and Child Neurology*, *37*, 906–914.

Kakai, B. M. (2011). *Challenges facing students in adapting to mathematics learning in secondary schools in Bungoma South District, Kenya* (Unpublished master's thesis). Kenatta University.

Kothari, C. R. (2004). *Research methodology: Methods and techniques*. New Age International Limited Publishers.

Land, S. M., & Hannafin, M. J. (2006). Student-centered learning environments. In D. H. Jonassen & S. M. Land (Eds.), *Theoretical foundations of learning environments* (pp. 1–23). Lawrence Erlbaum Associates, Inc.

Lerkkanen, M. K., Kiuru, N., Pakarinen, E., Viljaranta, J., Poikkeus, A. M., Rasku-Puttonen, H., Siekkinen, M., & Nurmi, J. E. (2012). The role of teaching practices in the development of children's interest in reading and mathematics in kindergarten. *Contemporary Educational Psychology*, *37*(4), 266–279. doi:10.1016/j.cedpsych.2011.03.004

Little, M. E. (2009). Teaching mathematics: Issues and solutions. *Teaching Exceptional Children Plus*, *6*(1), 12–22.

Maccini, P., Mulcahy, C. A., & Wilson, M. G. (2007). A follow-up of mathematics interventions for secondary students with learning disabilities. *Learning Disabilities Research & Practice*, *22*(1), 58–74. doi:10.1111/j.1540-5826.2007.00231.x

Mackay, C. B. (2001). *Republic of Kenya: Second University in Kenya, Report of Presidential working Party*. Government Printers.

McTighe, J., & O'Connor, K. (2006, Summer). Seven practices for effective learning. *Educational Leadership*, *63*(10), 13–19.

Merlot Pedagogy. (2013). *A program of the California State University in partnership with higher education institutions*. Author.

Miheso, M. M. (2002). *Factors affecting mathematics performance among Secondary Schools Students in Nairobi Province, Kenya* (Unpublished M.Ed. thesis). Kenyatta University.

Miller, S., & Mercer, C. (2017). The educational aspects of mathematical disabilities. *Journal of Learning Disabilities*, *30*(1), 47–56. doi:10.1177/002221949703000104 PMID:9009876

Murray, T., & Arroyo, I. (2002). *Towards measuring and maintaining the zone of proximal development*. Plenum Press.

Muvango, W. M. (2013). *Teachers have to be more creative*. Saturday Nation.

National Center for Education Statistics. (2004). *Highlights from the trends in International mathematics and science study (TIMSS)*. U.S. Government Printing Office.

Oyeniran, J. O. (2013). Teaching methods. An introduction to principles and methods of teaching. Lagos: SIBIS Ventures.

Piaget, J. (1952). *The child's concept of number*. Humanities Press.

Saunders, M., Lewis, P., & Thornhill, A. (2007). *Research methods for business students* (4th ed.). Prentice Hall.

Schramn, W. (1977). *Big media, little media, tools and techniques for instructions*. Sage Publications.

Stodolsky, S. S. (2008). *The subject matters*. University of Chicago Press.

The Report of the Expert Panel on Mathematics in Grades 4 to 6 in Ontario. (2004). *Teaching and Learning Mathematics*. Author.

Vygotsky, L. S. (1962). *Thought and Language*. MIT Press. doi:10.1037/11193-000

Vygotsky, L. S. (1978). *Mind in society: The Development of Higher Psychological Processes* (M. Cole, V. John-Steiner, S. Scribner, & E. Souberman, Eds.). Harvard University Press.

Wells, G. (2000). *Dialogic inquiry: Towards a socio-cultural practice and theory of education*. Cambridge University Press.

Chapter 8
Machine Learning Algorithms for Solving Linear Systems of Equations

Hilal Arslan
İzmir Bakircay University, Turkey

Fatma Bozyigit
İzmir Bakırçay University, Turkey

ABSTRACT

Moving into the fourth industrial revolution and the rapid digital transformation, there is a huge volume of data to be managed in each industry. Industrial simulations commonly produce data including the inputs and outputs of linear systems with several million unknowns. Solving linear systems is one of the fundamental problems in scientific computing, and it requires significant system resources. Determining a suitable method to solve linear systems can be a challenging task, since there is not a certain knowledge about which method is the most suitable for different numerical problems. In this study, the authors demonstrate how machine learning (ML) approach can be used in selecting solvers for linear systems. The chapter includes frequently used ML methods from literature and explain the usage of them to select optimal solvers and preconditioners.

DOI: 10.4018/978-1-7998-3868-5.ch008

INTRODUCTION

We are in the center of a significant transformation regarding the way production with the arrival digitization of manufacturing. This transformation is called Industry 4.0 to represent the fourth revolution that has occurred in manufacturing. In Industry 3.0, automation started with the adoption of computers. Industry 4.0 enhances it with smart and autonomous systems fed by data. Computers are connected and communicate to make decisions without human involvement ultimately. A combination of cyber-physical systems, the Internet of Things, and the Internet of Systems make Industry 4.0 possible and the smart factory a reality. With the support of intelligent machines getting smarter as they access to more data, factories will become more efficient and productive and less wasteful.

As moving into the Fourth Industrial Revolution with the rapidly digital transformation, there is a huge volume of data to be handled in each industry. Since connected machines create and share a tremendous volume of data, analyzing the data to identify patterns and insights would be impossible for a human to do in a sensible time. Industrial simulations generally produce data including the inputs and outputs of linear systems with several million unknowns. For example, the data in such domains (civil engineering, mechanical engineering, chemistry, biology, economics, etc.) arises as sparse matrices in which most of entries are zero. Determining a suitable method to solve these applications is a major problem requiring knowledge in the specific discipline like mathematics, and certain degree of expertise. Moreover, there is not a certain knowledge about which method the most suitable for many numerical problems is, or what the relationship between the kind of problem and the behavior of the solution methods.

Industrial data collected from sensors in its equipment or stored in the cloud environment can be in many forms such as sparse matrices as stated previous paragraph. Processing of these data requires some matrix computations and solution of the linear systems. Solution methods are generally grouped into two categories which are direct and iterative. Iterative methods are utilized for solving large sparse linear systems since they require less memory and operation count especially low accuracy is desired. Determining the best method to solve the linear systems through execution time and accuracy is challenging task. Up to date studies use machine learning (ML) techniques for identifying the best solver and preconditioner for a given problem. ML enable to process vast amounts of data to create predictive models. This supports researchers in developing classifiers which take a set of linear system attributes as input and returning preconditioned solver as the output.

Many real-world problems require solution of linear systems. There are several methods for solving such systems, and it is critical to choose the method that best suits the problem area. The state-of-the-art methods use machine learning methods for determining the best method for the problem solution. In this chapter, we review machine learning techniques for determining optimum linear solver and preconditioner for solving such systems. The chapter is organized as follows: In section I, we introduce the problem of the solution of the linear systems and application areas. In section II, some background information about preconditioning for solving the linear systems is given. In section III, existing linear solvers are explained and compared regarding their performance and complexities. Determining good set of features is a fundamental problem in forming good classifiers. In section IV, we overview domain specific features, sparsity pattern features, and generic features, which give important clues for understanding sparse matrices and provide guidelines for choosing appropriate linear solver and preconditioner. Finally, in section V, we survey the state-of-the-art machine learning approaches for choosing optimum linear solver and preconditioners.

Background

Artificial intelligence and machine learning methods have been identified as drivers of the Fourth Industrial Revolution. Machine learning methods can be efficiently used for choosing optimum method for solving the linear systems arising in many areas. Direct or iterative methods are preferred for solving linear systems. These methods have some parameters that effect their performance or that is able to benefit from data preprocessing or transformations such as preconditioning. The term preconditioning refers to transforming the system Ax = b into another system which is easily solved by using an iterative solver. The main idea behind the preconditioning is to improve the spectral properties of the coefficient matrix A. The choice of the preconditioner depends heavily on the properties of the coefficient matrix, and the type of iterative method conforming with. If M is a nonsingular matrix, then the linear system $M^{-1}Ax = M^{-1}b$ that can be easily solved, has the same solution as Ax = b. In this equation, M is called as the preconditioner which improves the spectral properties of the coefficient matrix. There are two main requirements for choosing a good preconditioner. First, solving the preconditioned system should be easier than the original system. Second, computation time for the preconditioned system should be less than that for the original system.

One of the most popular methods for preconditioning sparse matrices is the incomplete LU factorization (ILU), which is based on the LU factorization and attempts to find an approximation of L and U such that $A \approx LU$. It obtains similar results to the LU factorization using fewer arithmetic operations. Another well-known preconditioning technique is the incomplete Cholesky factorization. It is effectively used for solving positive definite linear systems arising in many areas such optimization problems.

Solution of Linear Systems

Systems of linear equations arise in a large number of areas in many scientific simulations (Arslan and Manguoglu, 2019), engineering applications problems (Cichocki et al., 1992), computational fluid dynamics (Benson, 1992), graph theory (Spielman, 2011), optimization (Arslan, 2019), circuit design (Zhang et al., 2011), medical imaging (Censor, 1988), and the shortest path applications (Arslan and Manguoglu, 2019). Thus, they constitute one of the most important branches of scientific computing.

A system of linear equations can be expressed in the simple form

$$Ax = b \tag{1}$$

where A is the coefficient matrix, b is the right hand side vector, and x is an unknown vector. The coefficient matrix of the linear systems can be grouped in two categories which are sparse or dense. A sparse matrix is defined as a matrix in which majority of the values are zero, and a dense matrix is a matrix containing very few non-zero elements. The solution methods of the linear systems are determined by sparsity of the matrix and investigated into two categories, direct and iterative.

Direct methods are based on factorization of the coefficient matrix A, and are widely used in many industrial codes in which the main concern is reliability. Moreover, they are preferred in certain areas including chemical plant modelling (Farina and Acattolini, 2012), fluid dynamics (Bolognani and Dörfler, 2015), circuit design (Cao et al., 2013), semiconductor device modelling (Reiser, 1972), and power system networks (Song et al., 2016). Direct methods are robust, and they require a predictable amount of time and storage. However, they have some disadvantages. They poorly scale with problem size with regards to memory and operation counts. When considering on large scale linear systems with several millions of unknowns, the only choice is to use iterative methods for solving such systems. Although iterative methods are less reliable than direct methods, they require fewer operation count and memory storage especially when the low accuracy is desired. Iterative methods are widely used in areas such as nuclear power industry (Tucker, 2015), and petroleum industry (Manne, 1958). In some applications, iterative methods may fail and preconditioning is necessary to improve the convergence in an acceptable amount of time and reliability.

Direct Methods

Direct methods are used to solve linear systems, and a method is called as a direct when the solution is computed within a finite number of operations using exact arithmetic. They are attractive due to their efficiency, reliability, and generality, and are generally preferred for solving the linear systems whose coefficient matrix is dense. In this part, widely used direct methods, Gaussian elimination, LU factorization, and Cholesky factorization are given in detail.

Gaussian Elimination

The main idea behind Gauss elimination method is to reduce the linear systems whose coefficient matrix is upper triangular by applying elementary row operations. After obtaining the linear systems whose coefficient matrix is upper triangular matrix, back substitution is used to find the solution. Back substitution contains isolating a single unknown variable, and substituting it back into another equation in order to isolate another unknown variable. The process continues until all unknowns are found. There are three elementary row operations, which can be used to transform the linear system to one whose coefficient matrix is upper triangular. The first one is row swapping, which exchanges two rows with each other. The second operation is row multiplication, which multiplies a row by a scalar value. The last one is row addition, which adds one row to another. Gaussian elimination may be given into three steps:

(1) The linear system is written as a matrix.
(2) The elementary row operations are used to reduce matrix to upper triangular one given in Algorithm 1.
(3) Back substitution is used to solve the linear system.

This process can be carried out in $O(n^3)$ floating point operations where n is the number of equations and given in Algorithm 1. Gaussian elimination is not efficient for large linear systems, but still commonly used in many fields. For instance, a mesh-connected multiprocessor structure is considered as the problem of solving a linear system, and the solution is obtained by using Gaussian Elimination (Masood, 2015). Another application area is scheduling, which is the most important problem in parallel processing, and the problem can be handled efficiently by using Gaussian Elimination (Srinivas, 1983). Error correction technique known as channel coding is used to avoid the errors for transmission of data encoded by using Gaussian Elimination (Atif and Rauf, 2009).

```
Algorithm 1 Gaussian Elimination (Baraiss, 1968)
for k=1 to n-1 do
for i = k+1 to n do
A(i,k) = A(i,k)/A(k,k)
for j = k+1 to n do
A(i,j) = A(i,j) - A(i,k)A(k,j)
```

LU Factorization Method

Let A be a nonsingular matrix. The LU factorization of A is represented by

$$A=LU \tag{2}$$

where L is a lower triangular matrix with ones on the diagonal, and U is an upper triangular matrix. Therefore, the linear system is easily solved by following two steps:

(1) $Ly = b$ is solved to find y using forward substitution.
(2) $Ux = y$ is solved to find x using backward substitution.

The *LU* factorization method can be effectively used to solve the linear systems with multiple right hand sides. The pseudocode of the algorithm is given in Algorithm 2. It can be computed in-place for dense matrices. That is, all entries in the *LU* factorization is able to be stored in the input matrix, which saves *U* values in the upper triangle of the matrix and saves L values under the diagonal since the main diagonal of *L* does not need to be saved. In this way, the use of addition memory is prevented. The floating point operations required by *LU* factorization of a square matrix is ($2n^3/3$). The method can be used for various areas including implementing fault tolerance methods to increase both reliability and resiliency (Coleman et al., 2017), and in many scientific simulations including circuit simulation (Lee, Achar, and Nakhla, 2018).

```
Algorithm 2 LU factorization (Saad, 1994)
for k=1 to n-1 do
U (k, k) = A (k, k)
for i=k+1 to n do
L (i, k) = A (i, k) / U (k, k)
U (k, i) = A (k, i)
for i=k+1 to n do
for j=k+1 to n do
A(i, j) = A(i, j) - L(i, k) × U (k, j)
```

Cholesky Factorization

If A is a symmetric positive definite matrix, then it can be decomposed

$$A = LL^T \tag{3}$$

where L is a lower triangular matrix with real and positive diagonal entries, and L^T is the conjugate transpose of L, which is called the Cholesky factorization. After the coefficient matrix is decomposed, then linear systems shown in Equation 4 and 5 are solved, respectively.

$$Ly = b \text{ for unknown y} \tag{4}$$

$$L^T x = y \text{. for unknown x} \tag{5}$$

The Cholesky factorization given in Algorithm 3 is considered to be amongst the most numerically stable of all matrix factorizations, and requires n*3*/3 floating point operations. The method can be used in many areas such as the Monte Carlo method (Bitar et al., 2017) and Kalman filtering (Nino-Ruiz, Sandu, and Deng, 2018).

Algorithm 3 Cholesky factorization (Higham, 2009)

1: **for** j=1 to n **do**

2: $\mathbf{L}(\mathbf{j},\mathbf{j}) = \left(\mathbf{A}(\mathbf{j},\mathbf{j}) - \sum_{k=1}^{j-1} L(j,k) L(j,k) \right)^{1/2}$

3: **for** i=j+1 to n **do**: $\mathbf{L}(\mathbf{i},\mathbf{j}) = \left(\mathbf{A}(\mathbf{i},\mathbf{j}) - \sum_{k=1}^{j-1} \mathbf{L}(\mathbf{i},\mathbf{k}) \mathbf{L}(\mathbf{j},\mathbf{j}) \right) / \mathbf{L}(\mathbf{j},\mathbf{j})$

Iterative Methods

Iterative methods are used to obtain an approximate solutions of certain classes of the linear systems. They start with an initial guess of the solution and improve the initial guess in a finite sequence by replacing Equation 1 with an iterative system of the form

$$x_{k+1} = Tx_k + c \text{, k =1, 2,...} \tag{6}$$

where T is an n by n iteration matrix, and c is a scalar value. Passing from one iteration to the next one is included by changing one or few components of an approximation to the solution at a time. Thus, the solution is obtained by improving iterations. Iterative methods of solving the linear systems have some advantages over elimination methods. Elimination methods are prone to rounding errors. Moreover, they could consume much memory due to fill-ins. Furthermore, they require large number of arithmetic operations and memory especially when the coefficient matrix is large. For these reasons, iterative methods are generally preferred for solving large linear systems. In this part, an overview of the iterative methods, Jacobi, Gauss-Seidel, and Successive Over-Relaxation is given. In these iterative methods, the coefficient matrix A is divided by

$$A = D+L+U \tag{7}$$

where D is the diagonal of A, L is the lower triangular part of A, and U is the upper triangular part of A.

Jacobi Method

The Jacobi method is the first iterative method for solving the linear systems can be expressed as

$$x_{k+1} = -D^{-1}\left(L+U\right)x_k + D^{-1}b. \tag{8}$$

The pseudocode of the algorithm is given in Algorithm 4. Jacobi method can be effectively used for solving equations arising from many fields such as signal processing, chemical physics, statistics, control theory of dynamical systems, electrical networks, electrochemistry of corrosion, probability, and statistics (Doa, Bhrawy, and ezz-Eldien, 2012).

Algorithm 4 Jacobi Method (Rutishauser, 1966)

```
for k=1, … do
for i=1 to n do
x(i) = 0
for j=1, 2,...,i-1,i+1,…,n do
x(i)=x(i)+A(i,j)x^(k-1)(j)
x(i)=(b(i)-x(i))/A(i,j)
x^(k) = x
check convergence; continue if necessary
```

Gauss-Seidel Method

The Gauss-Seidel method is an iterative method for solving Equation 1, and has also been considered in many applications such as inverse modelling (Ng, Boroujeny, and Garg, 2001), speech processing, control system (Widrow and Stearns, 1985), and real-time optic flow computation (Bruhn et al., 2003) mainly because of its ease of implementation and small storage requirement. Gauss-Seidel method relies on a refinement implementation of the Jacobi method, which is expressed as

$$x_{k+1} = Cx_k + f \tag{9}$$

where $C = -(I + D^{-1}L)^{-1} D^{-1}U$ and $f = (I + D^{-1}L)^{-1} D^{-1}b$,and given in Algorithm 5. When the coefficient matrix of the linear system is either diagonally dominant or symmetric, and positive definite, the method guarantees convergence.

Algorithm 5 Gauss-Seidel Method (Kearfott, 1990)

```
for k=1, ... do
for i=1 to n do
r = 0
for j=1, 2,..., i-1 do
r =r + A (i, j) x^(k-1) (j)
for j=i+1,..., n do
r =r+A (i, j) x^(k) (j)
x^(k) (i) = (b(i) - r)/A (i, i)
```

Successive Over-Relaxation Method

Designing new methods for accelerating the convergence to solution is one of the goals in numerical methods. The successive over-relaxation method (SOR) is discovered by making a simple modification of the Jacobi and Gauss-Seidel methods to improve convergence. The method is obtained by using a relaxation technique performed to the basic Gauss-Seidel iteration,

$$\left(I + wD^{-1}L\right)x_k + 1 = \left(\left(1 - w\right)I - wD^{-1}U\right)x_k + wD^{-1}b \tag{10}$$

where w is the relaxation parameter, and if w is set to 1, the method turns to the Gauss-Seidel. When the coefficient matrix is symmetric positive definite and

$0 < w < 1$, the method guarantees convergence. The optimum value of w is given by

$$w_b = \frac{2}{1+\sqrt{1-\mu_1^2}} \tag{11}$$

where $0 < \mu_1 < 1$ is the largest eigenvalue of the Jacobi iteration matrix, $D^{-1}\left(L+U\right)$. A detailed estimation procedure of value w is found in (Hageman and Young, 1981). The method is used especially in the petroleum industry (Benzi, 2002), fluid dynamics, mathematical programming, linear elasticity, and heat conduction (Mayooran and Light, 2016).

Krylov Subspace Methods

We have formed two classes of algorithms for solving linear systems, so far. The first one is *direct methods* which are based on matrix factorization and compute the exact solution. The second one is *iterative methods* and they approximately compute the solution. Krylov subspace methods introduced by Alexei Krylov combine direct and iterative methods, and they are referred by *semi-direct methods* which converge the exact solution in a finite number of steps. Krylov subspace methods have become popular recently due to their simplicity and generality. There are various Krylov based methods that are iteratively compute an approximation of the exact solution x by transforming an n-dimensional vector space to a lower m-dimensional subspace using matrix-vector multiplications instead of estimating A^{-1}. Let x_0 be an initial guess, and the exact solution of the linear system be $x = x_0 + x_m$. The estimate of x_m forms an approximation in the Krylov subspace K_m :

$$K_m\left(A, r_0\right) = span\left\{r_0, Ar_0, \ldots, A^m r_0\right\} \tag{12}$$

where $r_0 = b - Ax_0$. Krylov subspaces start with a single vector space, and add one additional vector at a time. Thus, an accurate approximation to the solution can be obtained on a relatively small dimension of a search space. Different kind of Krylov subspace methods emerge from different choices of the subspace K_m. A brief overview of the methods based on Krylov subspaces is given in the following.

Conjugate Gradient Method

The Conjugate Gradient method (CG) is the current iterative method for solving the linear systems whose coefficient matrix is positive definite. It is remarkably faster than other methods like SOR and does not require any parameter tuning as well as will provide the solution in at most n steps. Given A, b, an initial estimate to the solution x_0, and a tolerance, the main steps of the CG method can be outlined in Algorithm 6. The method is effectively used in many fields including robotics (Sun, Tian, and Wang, 2017), and large scale optimization problems (Andrei, 2008).

Algorithm 6 Conjugate Gradient Method (Powell, 1977)

```
r_0 = b-Ax_0
p_0 = r_0
for i=1,…,numiter do
α_{i-1} = (r^T_{i-1} r_{i-1})/(p^T_{i-1}Ap_{i-1})
x_i = x_{i-1} + α_{i-1} p_{i-1}
r_i = r_{i-1} - α_{i-1} Ap_{i-1}
if ||r_i|| < tolerance then
return
β_i = (r^T_i r_i)/(r^T_{i-1} r_{i-1})
p_i = r_i + β_i p_{i-1}
```

The conjugate gradient method is not applicable for unsymmetrical systems. In this case, the biconjugate gradient method (BiCG) may be used. BiCG replaces the orthogonal residual sequences to mutually orthogonal sequences, which do not provide a minimization. To update the residual, BiCG uses the dual relation A^T instead of A. There are various alternatives of BiCG method. One of the alternatives is the biconjugate gradient stabilized method (BiCGStab) used for solving general linear systems whose coefficient matrix is asymmetric. When comparing to the BiCG, BiCGStab often converges twice as fast as the BiCG. BiCGStab locally minimizes the residual vector, which results in a considerably smoother convergence behavior. Given A and b, and initial guess x_0, the main steps of BiCSTab method is presented in Algorithm 7. The method is widely used in biomedical imaging and geoscience (Zhang et al., 32).

Algorithm 7 BiCGStab Method (Gutknesht, 1993)

```
r_0 = b - Ax_0 .
p_0 = r_0
for j= 0, 1, … do
```

$\tilde{p}_j = M^{-1} p_j$

$\alpha_j = (r_j, \bar{r}_0) / (A\tilde{p}_j, \bar{r}_0)$

$s_j = r_j - \alpha_j A\tilde{p}_j$

$\tilde{s}_j = M^{-1} s_j$

$\omega_j = (A\tilde{s}_j, s_j) / (A\tilde{s}_j, A\tilde{s}_j)$

$x_{j+1} = x_j + \alpha_j \tilde{p}_j + \omega_j \tilde{s}_j$

$r_{j+1} = s_j - \omega_j A\tilde{s}_j$

$\beta_j = (r_{j+1}, \tilde{r}_0) / (r_j, \tilde{r}_0) . \alpha_j / \omega_j$

$p_{j+1} = r_{j+1} + \beta_j (p_j - \omega_j A\tilde{p}_j)$

Generalized Minimum-Residual Method

Generalized Minimum-Residual method (GMRES) is an iterative method for solving the linear systems whose coefficient matrix is asymmetric. The method is based on the residual minimization over a Krylov subspace, and consists of two main steps. In the first step, an orthonormal basis is created by the Arnoldi process (Mantinfar, 2011). In the second step, a least square problem is solved to change the last approximation by using the orthonormal basis vectors. In this step, GMRES starts to an initial guess x_0 and forms the kth iteration as $x_k = x_0 + z_k$ where x_k is chosen so that the norm of corresponding residual r_k is minimized over $x_0 + K_k(r_0)$. That is,

$$\| r_k \| = \| r_0 - Az_k \| = \min_{z \in x_0 + K_k(r_0)} \| r_0 - Az \| \tag{13}$$

where $r_0 = b - Ax_0$ and $K_k(v) = span\{v, Av, \ldots, A^{k-1}v\}$. The main steps of GMRES method are presented in Algorithm 8. Although it has higher operation counts and memory requirement than other iterative methods, it is among the most robust methods for solving nonsymmetrical systems. GMRES is used in many applications such as aviation industry (Nakada, 2018), seismic imaging methods such as the full wave inversion (Virieux and Operto, 2009).

Algorithm 8 Generalized Minimum Residual Method (Saad and Schultz, 1986)

$r_0 = b - Ax_0$

2: v1 = r0/$||r_0||_2$
3: create (k+1) orthogonal vectors $v_1, v_2, \ldots, v_{k+1}$ as a basis for $K_{k+1}(r_0)$
4: find $\tilde{y} \epsilon \boldsymbol{R}^k$ as solution of least square problem (Equation 13)
5: update $\boldsymbol{x}_k = \boldsymbol{x}_0 + \boldsymbol{V}_k \tilde{\boldsymbol{y}}$ if x_k does not satisfy the termination condition,
Set x_0 to x_k and go to 1.

There is an extension of GMRES, called as Flexible GMRES (FGMRES) allowing larger flexibility in the choice of solution subspace than GMRES. FGMRES is widely used to solve ill-posed problems which GMRES does not always perform well. Another Krylov subspace method with many similarities to the GMRES method is the transpose-free quasi-minimal residual method (TFQMR). The term Quasi-minimal implies the produced Krylov subspace being bi-orthogonal instead of orthogonal like in the GMRES.

Machine Learning Methods to Select Optimal Solvers and Preconditioners for Linear Systems

As moving into the Fourth Industrial Revolution with the rapidly digital transformation, there is a huge volume of data to be handled in each industry. Industrial simulations generally produce data including the inputs and outputs of linear systems with several million unknowns. Determining a suitable method to solve these applications (see Figure 1) is a major problem requiring knowledge in the specific discipline like mathematics, and certain degree of expertise. Moreover, there is not a certain knowledge about which method the most suitable for many numerical problems is, or what the relationship between the kind of problem and the behavior of the solution methods.

Since many users cannot be enough knowledge about tuning solver parameters or preconditioning options, many numerical libraries for solving scientific problems establish a conservative solver as default. Although a default solver may be good option for many problems, it cannot work or converge for some problems, or not much faster than other methods. To improve the efficiency of large scale simulations in science and engineering, machine learning methods are used for determining optimum solvers by revealing the particular properties of different input problems.

Figure 1. General framework of determination of optimal solver

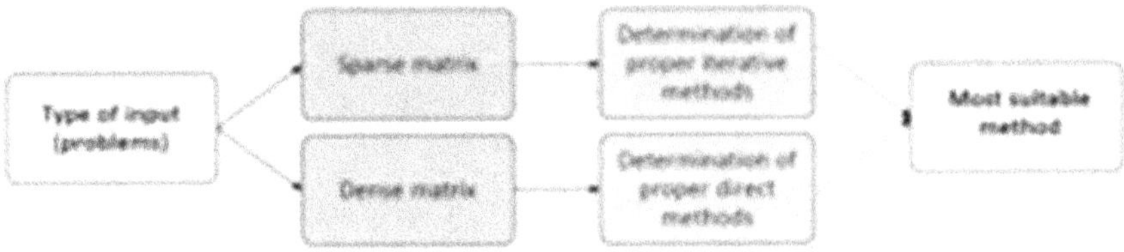

General Framework of ML Classifier

There have been several investigations and software developments related to determining efficient solvers for linear systems. In this study we review solver selection studies using ML algorithms. These algorithms consist of two basic parts such as feature selection and classification (see Figure 2). In the feature selection part, necessary features of the instances in the dataset are specified and redundant ones eliminated. Selected features give information about characteristics of the system and increase the performance of solvers. The classifier then maps the given feature sets to a choice of solver or preconditioner for the linear systems.

Figure 2. General framework of a classifier

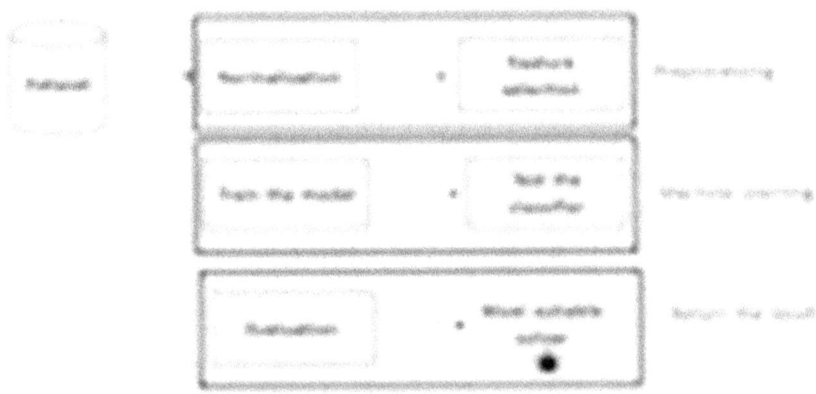

Feature Selection

As there is a huge volume of data to be handled in each industry, the presence of ML makes productive use of data. Data science techniques using ML fundamentals help almost all industries to reap in benefits and helps companies to get meaningful knowledge from historical data. Accordingly, the companies plan business strategies to make the existing source more productively efficient.

It is unquestionable that using appropriate and informative features is a crucial step for the classification process. Features are explained by their generality and computational cost in the following categories. The candidate set of features can be presented in Table 1. Basic features describe sparsity structure of the matrix, such as the number of nonzero elements per row, bandwidth, and the number of structurally asymmetric elements, 1-norm, infinity-norm, and frobenius-norm of the matrix. Computing such features has low complexity. Advanced features describe the spectrum of the matrix. Computation of these features is expensive, so they are computed approximately. These features include spectral radius, eigenvalues, condition number, and singular values. Domain-specific features include mesh properties. Domain specific knowledge is able to supply better insights to convergence of the linear systems.

Table 1. List of candidate features for solver prediction. They are investigated into three groups which are basic, advanced, and domain specific features (Bhowmick et al., 2011).

Basic Features
number of rows
number of non-zeros
number of non-zeros in lower triangle
number of non-zeros in upper triangle
maximum number of non-zeros per row
average number of non-zeros per row
largest element in magnitude along diagonal
smallest non-zero element in magnitude along diagonal
bandwidth of lower triangle
bandwidth of upper triangle
number of ones
number of rows that are strictly diagonally dominant, i.e., satisfying $2*\left\lvert a_{j,j}\right\rvert - \sum_{j}\left\lvert a_{i,j}\right\rvert > 0$
normalized sum of distances from diagonal of non-zero elements in lower triangle $\sum(i-j)/S$ where $i > j$, $a_{i,j} \neq 0$, and $S = \sum_{i=1}^{N_{row}-1} i*(N_{row}-1)$

continued on following page

Table 1. Continued

Basic Features
normalized sum of distances from diagonal of non-zero elements in upper triangle $\sum(j-i)/S$ where $j>i$, $a_{i,j} \neq 0$, and $S=\sum_{i=1}^{N_{row}-1} i*(N_{row}-1)$
property showing how symmetric the matrix is as 1- $nnz(A-A.^T)/(nnz(A)-nnz_d(A))$ where $A.^T$ is a non-conjugate transpose of A and $nnz_d(A)$ is number of non-zeros on diagonal of A
property showing how skew-symmetric the matrix is as 1- $(nnz(A+A.^T)-nnz_d(A))/(nnz(A)-nnz_d(A))$
property showing how symmetric the non-zero pattern of matrix is as $1-nnz(P-P.^T)/(nnz(P)-nnz_d(P))$ where $p_{i,j}=1$ if $a_{i,j}\neq 0$, 0 otherwise
1-norm
Frobenius norm
Advanced Features
spectral radius
second largest absolute eigenvalue
non-zero smallest absolute eigenvalue
where $\lambda_1=\rho(A)$
$peak=\lambda_1/((\sum_i \lambda_i)-\lambda_1)$
where κ is the condition number computed by Matlab
portion of $\sqrt{\sigma}>0.9\text{x}\left(\sqrt{\sigma_{max}}-\sqrt{\sigma_{min}}\right)+\sqrt{\sigma_{min}}$, where σ is a singular value
$1-(s_{90}+s_{10})$
portion of $\sqrt{\sigma}\leq 0.1\text{x}\left(\sqrt{\sigma_{max}}-\sqrt{\sigma_{min}}\right)+\sqrt{\sigma_{min}}$
Domain Specific Features
number of spatial dimensions (from the mesh)
N_{row}/N_{dim} if P2, N_{row} otherwise
finite element discretization order
mesh refinement level
approximation to condition number term $\sqrt{\kappa}$, where $\kappa'=\left(N'_{row}\right)^{2/N_{dim}}$
approximation to condition number term $\sqrt{\kappa}$, where $\kappa'_{hmin}=\left(1/\min(h)\right)^2$
approximation to condition number term $\sqrt{\kappa}$, where $\kappa'_{hmax}=\left(1/\max(h)\right)^2$

Feature selection is a pre-processing step in ML to determine features which contribute most to be predicted output by excluding irrelevant data columns in the dataset. A basic feature selection process comprises of generating a subset of the features, determination of its contributions to classification results.

Feature selection must be separated from modelling ML classifiers. Most algorithms for learning classifiers underperform on the test set when the number of features exceeds the size of the training set. Therefore, the feature subset must be determined carefully. Specification of optimal features is a challenging task since it depends on the number of samples in the training set. Features that are meaningful when the training set is large often, otherwise they have an adverse impact on the test set performance.

Modelling Classifier

After pre-processing procedure, the data is divided into training and test sets. The chosen ML algorithm is trained on the previously-selected subset of features (train set) and the model evaluated according to classification results using test. The accuracy of the classifier is measured by using different evaluation metrics such as precision, recall, F-measure, mean square error (MSE), and so on.

Since the choice of solver is basically related to matrix characteristic, pre-trained ML model is primarily focuses on specification of the sparsity. Then, the solvers that previously perform well for systems which have similar matrix data with the target problem are suggested as applicable.

Baseline Machine Learning Algorithms Experimented in Related Domain

There are many researches in the literature that experiment with the different methods to determine optimum linear solver and preconditioner for solving such systems. Recently, ML has become one of the widely used information processing approach to determine linear solvers. Figure 3 illustrates the most preferred baseline ML learning algorithms in this framework. Evaluation results of the applied methods show that baseline ML approaches generally outperform the frequently used methods from the literature.

Figure 3. Baseline ML learning algorithms used for selecting optimal solvers and preconditioners

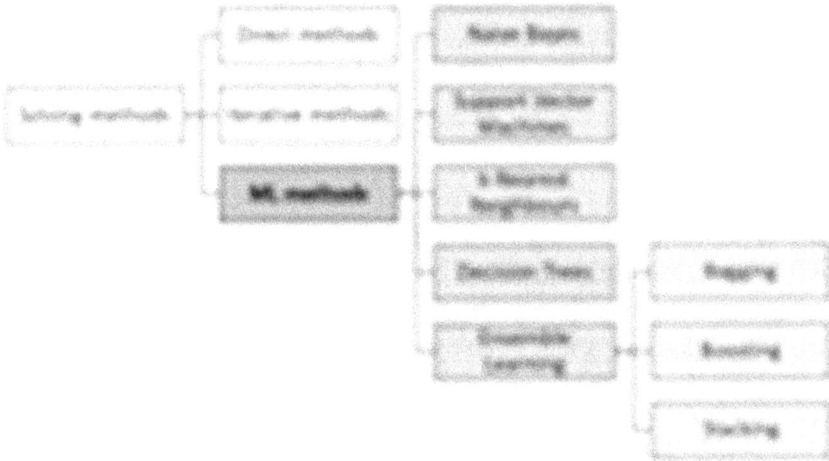

Naïve Bayes (NB)

NB is a frequently used ML classification algorithm based on Bayes' theorem which provides evaluation of explicit probabilities for any hypothesis. The theorem states that:

$$P(h|D) = \frac{P(D|h)P(h)}{P(D)} \tag{14}$$

where $P(h)$ denotes prior probability of hypothesis h, $P(D)$ is prior probability of training data D, $P(D|h)$ is probability of D given h, and $P(h|D)$ is probability of h given D. In order to choose the acceptable hypothesis the most probable one is selected.

NB classifier is based on the simplifying assumption that the attributes are conditionally independent given target value (Bozyiğit et al., 2019). In other words, the NB classifier assigns the most likely class to a given example described by its feature vector. Learning such classifiers can be greatly simplified by assuming that features are independent given class, that is,

$$P(X|C) = \prod_{i=1}^{N} P(X_i|C) \tag{15}$$

where $X = (X_{1,} \ldots, X_N)$ states a feature vector and C is a class.

NB is easy to implement algorithm and it is widely used in classification problems. Recently, application of NB classifier to choose an optimal linear solvers and preconditioners, has been widely conducted and presented. For example, Motter et al. (2017) develop a tool called Lighthouse and focus on its capabilities in the area of parallel iterative solvers for sparse linear systems They examine several ML methods NB, kNN, SVM, and Decision Tree methods for measuring accuracy and performance for the Trilinos library (Heroux, 2015) which is used in a variety of scientific applications, mainly focused around multiphysics simulations. They compare the performance of experimented ML methods to select the one that produces the best (most accurate) solver classification. They conclude that NB produce accuracy of 87.6% with sixty-eight input features among all ML methods they applied. In another study, Eijkhout and Fuentes (2008) use Decision Tree and NB classifiers to develop an iterative solver recommender for linear systems of equations.

Support Vector Machines (SVM)

In SVM method, an unlimited number of hyper planes are created to group the samples in the dataset and the most suitable one of these is selected. SVM is a supervised ML algorithm based on statistical information theory and basic risk minimization. In the first phase of SVM, an unlimited number of hyper planes are created to group the samples in the dataset. The separating hyper plane is created to be maximizing the distance between the plane and the nearest data points from both classes (Özçift, Kılınç, and Bozyiğit, 2019). The data points are called support vectors. Distance between the separating hyper plane and the support vectors is called a margin of the SVM. The area around the hyper plane can be increased before hitting any data point (see Figure 4). If there are points of one class cross over the hyper-plane and target classes are overlapping, the technique tries to minimize the distance of these misclassified points to their related location. SVM is a linear classifier because it divides a feature space into separate two plane. However, it can be used for non-linear classification problem by transforming the original feature space (Aiserman, Braverman, and Rozonoer, 1964). Although SVM can cope with over-fitting and performs well in case clear margin of separation between classes, it does not manage noisy data challenge. It also underperforms in cases where number of features exceeds the number of samples in training set.

Figure 4. Support vector machine hyperplane illustration lane

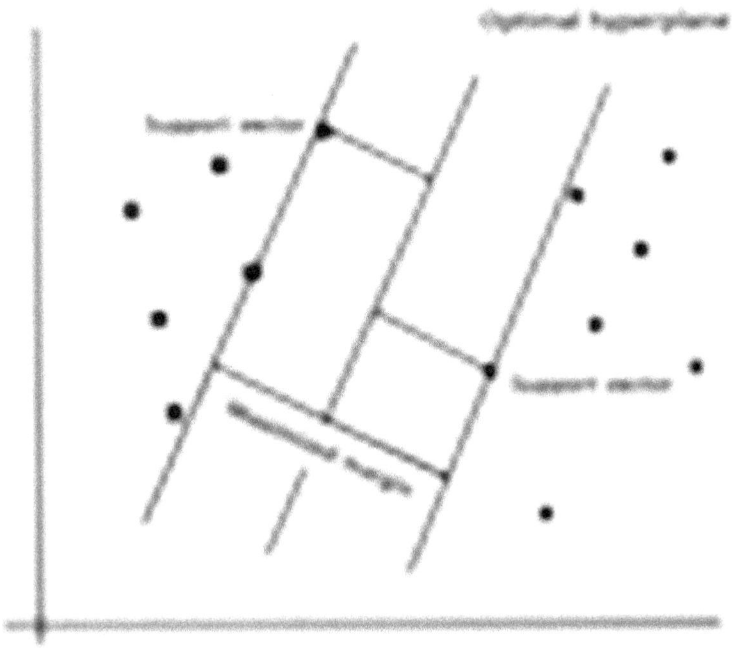

As SVM handle over fitting, it is one of the most preferable ML approach in the field of solving linear systems (selection of optimal linear solvers and preconditioners). For example, Bhowmick et al. (2011) demonstrate how ML techniques can be used in selecting linear solvers by experimenting SVM, NB, kNN, and Decision Tree. They state that SVM, NB, and kNN (for k=1 and k=10) achieve the best accuracies comparing the Decision Tree methods. They also claim that a set of only nine features (trace, diagonal-variance, diagonal average, number of rows, norm1, diagonal-dominance, number of zeros, left-bandwidth, and the condition number) is sufficient to create a classifier whose accuracy is comparable the accuracy of the classifier constructed using all features. The other studies using SVM to select optimal linear solvers are (Motter, 2017) and (Sood and Norris, 2017).

k Nearest Neighbor (kNN)

kNN is an instance-based classification algorithm which does not have a training phase (Kılınç et al., 2017). In the kNN algorithm, the input consists of the k closest instances with certain tags in the feature space. The distance between the samples can be calculated using different metrics such as Euclidean, Manhattan, Minkowski, and Hamming distances. The advantages of this method are that there is no training phase and it works well even with noisy data.

Work flow of kNN algorithm can be defined with the following steps;

1. The value of k (the nearest sample data) is specified.
2. Distance between test and each training data is calculated using any of the metrics as stated in previous paragraph.
3. k nearest instances are determined.
4. Test data is placed into a class regarding the most frequent class of k nearest neighbours.

kNN is mainly used for classification predictive problems in industry. Consequently, we can see many researches utilizing this ML method to predict solvers of linear systems. For example, Motter et al. (2015) evaluate kNN algorithm on the data generated by transient simulations to find optimal sparse linear solvers. In another study, Motter (2017) applies kNN on dataset including 1,586 sparse matrix obtained from the University of Florida's Sparse Matrix Collection for the selection of optimal iterative linear solvers. He states that perform better than other experimented ML methods (Gradient Boosting, NB, and AdaBoost). Moreover, Bhowmick et al (2011) observe the performance of kNN classifier (for k=1 and k=10) on a collection of fifty symmetric matrices, obtained from the University of Florida Sparse Matrix Collection (Davis, 1997). In the evaluation section of the study, researchers indicate that kNN classifier produces an accuracy over 84%.

Decision Tree

Decision Tree is a supervised learning algorithm that generates a tree using depth-first strategy to perform the classification task. The tree consists of nodes for attributes, branches for attribute values, and leaves for class labels (see Figure 5). The two basic steps in this technique are: building the tree based on information created using past experiences and modeling the tree through these data. Decision trees are especially practical for problems where samples in the dataset have many features. For instance, for the type of numerical problems on which this research is centered, there are around fifty different numerical features.

The construction of the decision tree structure is performed using training data. The class labels are located into leaf nodes by filtering each observation in the dataset through the tree structure.

A decision tree model bases on simple comparison rules actually. These rules are executed during the traversing time of a sample on the tree, features' values of the sample are compared with the contents of the visited node. The tree structure is formed recursively until all the samples in the training set reach the leaf nodes and are classified in a correct manner.

There are many different algorithms for constructing decision trees such as C4.5, CART, and so on. They differ in building the tree structure and handled data types generally.

Figure 5. Baseline ML learning algorithms used for selecting optimal solvers and preconditioners

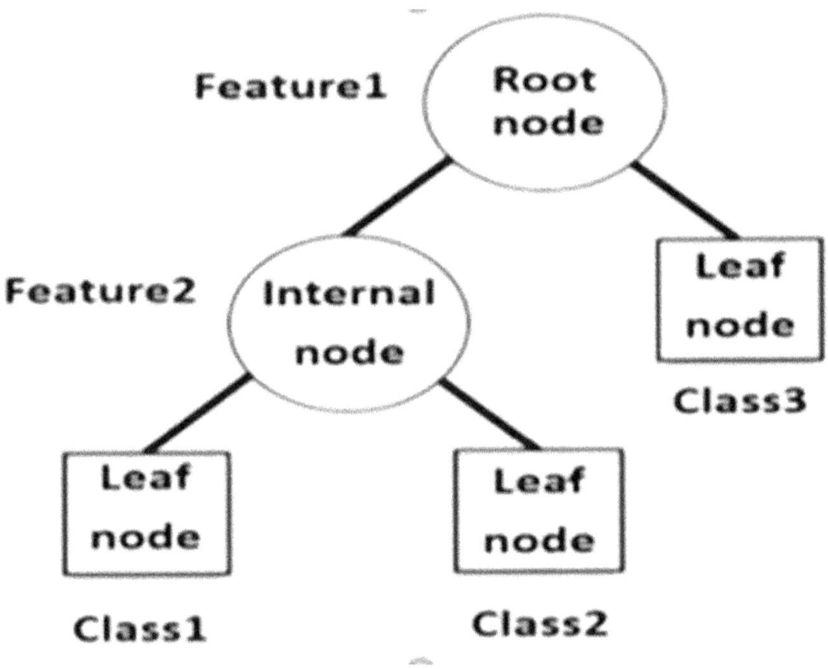

Decision Tree is one of the most preferable ML approach in the field of in selection of sparse linear solvers. For example, Bhowmick et al. (2011) utilize a ML technique, Alternating Decision Trees (ADT), to select efficient solvers based on the properties of sparse linear systems and runtime-dependent features, such as the stages of simulation. They present the empirical results over linear system obtained from computational fluid dynamics and hydrodynamics applications. The experimental results of the study state that ADT can deal with over-fitting. In another study, Motter (2017) performs Decision Tree, NB, kNN, Random Forest (RF), Gradient Boosting, and AdaBoost ML approaches on the matrices obtained from the University of Florida's Matrix Collection for the selection of optimal iterative linear solvers. He indicates that Decision Tree performs better than NB, Gradient Boosting, and AdaBoost. Eller et al. (2012) also conduct a ML based study for dynamic linear solver selection for transient simulations using C4.5 decision tree model. The researchers create a new dataset including sixty-five preconditioned solvers to observe classification results. Salın, et al. (2016) evaluate different approaches on a large set of matrices obtained from the University of Florida Sparse Matrix Collection. Experimental results of the study show that tree based classifiers (PART and J48) outperform the other classifiers by achieving more than 90% accuracy for five domains; directed graphs, undirected weighted graphs, economics problems, subsequent circuit simulation problems, and structural problems.

C4.5

C4.5 creates a decision tree on the current training set and estimates which class the input data belongs to Bozyiğit, Utku and Nasiboğlu, 2019). This method forms a decision tree structure by selecting the features which are the most discriminative. The advantages are that it can identify distinctive features, makes inference for estimation, and succeed in training sets with lost data.

CART

CART is widely used decision tree model which generates a binary tree. It uses an entropy model to select the most discriminative features in the dataset (Bittencourt and Clark, 2003). This approach tries to minimize the number of comparisons by ordering and grouping features that contribute the most information. An important characteristic of the CART method is the ability to manage over fitting by excluding certain nodes and branches to improve the generalization skills of the tree structure.

Ensemble Learning

Ensemble learning is a supervised ML paradigm where multiple models are trained to solve the same problem and combined to improve overall performance by decreasing variance, bias and improve the classification.

Ensemble methods can be examined in two categories; one is called sequential and the other is called parallel. The underlying idea of sequential methods is to use dependencies between the base learners. The overall performance can be improved (boosted) by reweighing previously mislabeled examples having bigger weight. The underlying idea of parallel methods is to use independencies between the base learners because the error can be significantly decreased by averaging. When we review the studies related to our research domain, we realize that only Motter (2017) experiments Ensemble Learning methods (RF, Gradient Boosting, and AdaBoost) on the matrices obtained from the University of Florida's Matrix Collection for the selection of optimal iterative linear solvers. He concludes that EL methods do not achieve accuracy results compared to NB, KNN, SVM and Decision Tree classifier.

Boosting

Boosting is an iterative approach to generate a robust classifier which sets the weight of samples based on the last classification. If an instance is misclassified, the model reweighs this observation. Boosting is a sequential ensemble learning method in which the first classifier is trained with random subset of the data set and the second is developed by fitting the residuals of the first. The third classifier is trained with instances on which first and second dissent. Finally, these classifiers are combined using simple majority voting.

Boosting decreases bias error and builds robust predictive models, but it tends to overfit the training data as well. Hence, parameter tuning is a critical process of boosting algorithms to deal with over fitting.

Random Forest (RF)

In this approach, first of all, the features of the samples are randomly selected during decision trees are created (Yıldırım and Birant, 2018). Then, the training data is designed to form each decision tree. Finally, the RF is created by bringing all the trees together. The classification process is carried out by voting of the trees in the RF and the class with the most votes is returned as a result. This classifier performs well on huge volume data.

CONCLUSION

Linear systems arise in many applications ranging from industrial problems to problems in numerical simulations and computations. The issue of selecting an efficient method to solve such problems is a significant but a challenging problem. Although there exist various numerical techniques for solving the linear systems in the literature, it seems that there are a few study using ML approaches to automate solution. In this study, we review and evaluate the ML based linear solver selection studies with respect to proposed approaches, functionalities, and the used dataset.

The analysis of previous studies reveals that there are some issues to be addressed in the literature. For example, current approaches generally apply baseline ML classification methods such as NB, SVM KNN, decision trees, and gradient boosting using WEKA tool to determine optimal linear solvers and preconditioners. Considering the main advantages of Neural Networks and Deep Learning architecture in prediction, planning, and uncertainty analysis, we believe experimenting of these methods will make significant contributions to literature.

REFERENCES

Aiserman, M. A., Braverman, E. M., & Rozonoer, L. I. (1964). Theoretical foundations of the potential function method in pattern recognition. *Avtomatika i Telemehanika*, *25*, 917–936.

Andrei, N. (2008). Another hybrid conjugate gradient algorithm for unconstrained optimization. *Numerical Algorithms*, *47*(2), 143–156. doi:10.100711075-007-9152-9

Arslan, H. (2019). Dynamic Physarum Solver: A bio-inspired shortest path method of dynamically changing graphs. *Turkish Journal of Electrical and Computer Science*, *27*(2), 1489–1503. doi:10.3906/elk-1811-46

Arslan, H., & Manguoglu, M. (2019). A parallel bio-inspried shortest path algorithm. *Computing*, *101*(8), 969–988. doi:10.100700607-018-0621-x

Arslan, H., & Manguoglu, M. (2019). A hybrid single-source shortest path algorithm. *Turkish Journal of Electrical and Computer Science*, *27*(4), 2636–2647. doi:10.3906/elk-1901-23

Atif, M., & Rauf, A. (2009). Efficient implementation of Gaussian Elimination method to recover generator polynomials of convolutional codes. *IEEE International Conference on Emerging Technologies (ICET 2009)*. 10.1109/ICET.2009.5353183

Bareiss, H. E. (1968). Sylvester's identity and multistep integer-preserving Gaussian elimination. *Mathematics of Computation*, *22*, 565–578.

Benson, D. J. (1992). Computational methods in Lagrangian and Eulerian hydrocodes. *Computer Methods in Applied Mechanics and Engineering*, *99*(2-3), 235–394. doi:10.1016/0045-7825(92)90042-I

Benzi, M. (2002). Preconditioning techniques for large systems: A Survey. *Journal of Computational Physics*, *182*(2), 418–477. doi:10.1006/jcph.2002.7176

Bhowmick, S., Eijkhout, V., Freund, Y., Fuentes, E., & Keyes, D. (2011). Application of alternating decision trees in selecting sparse linear solvers. In *Software Automatic Tuning* (pp. 153–173). Springer. doi:10.1007/978-1-4419-6935-4_10

Bitar, A. W., Ovarlez, J., & Cheong, L. (2017). Sparsity based Cholesky factorization and its application to hyperspectral anomaly detection. *2017 IEEE 7th International Workshop on Computational Advances in Multi-Sensor Adaptive Processing (CAMSAP),* 1-5.

Bittencourt, H. R., & Clarke, R. T. (2003, July). Use of classification and regression trees (CART) to classify remotely-sensed digital images. In *IGARSS 2003. 2003 IEEE International Geoscience and Remote Sensing Symposium. Proceedings* (IEEE Cat. No. 03CH37477) (Vol. 6, pp. 3751-3753). IEEE.

Bolognani, S., & Dörfler, F. (2015). Fast power system analysis via implicit linearization of the power flow manifold. *2015 53rd Annual Allerton Conference on Communication, Control, and Computing (Allerton)*, 402-409.

Bozyiğit, A., Utku, S., & Nasiboğlu, E. (2019, September). Cyberbullying detection by using artificial neural network models. In *2019 4th International Conference on Computer Science and Engineering (UBMK)* (pp. 520-524). IEEE. 10.1109/UBMK.2019.8907118

Bruhn, A., Weickert, J., Feddern, C., Kohlberger, T., & Schnörr, C. (2003). Real-time optic flow computation with variational methods. In N. Petkov & M. A. Westenberg (Eds.), Lecture Notes in Computer Science: Vol. 2756. *Computer Analysis of Images and Patterns. CAIP 2003*. Springer. doi:10.1007/978-3-540-45179-2_28

Cao, Y., Papageorgiou, A., Petras, I., Traub, J., & Kais, S. (2013). Quantum algorithm and circuit design solving the Poisson equation. *New Journal of Physics, 15*(1), 013021. doi:10.1088/1367-2630/15/1/013021

Censor, Y. (1988). Parallel applications of block-iterative methods in medical imaging and radiation therapy. *Mathematical Programming, 42*(1-3), 307–325. doi:10.1007/BF01589408

Cichocki, A., & Unbehauen, R. (1992). Neural networks for solving systems of linear equations and related problems. *IEEE Transaction on Circuits and Systems-Fundamental Theory and Applications, 39*(2), 124-138.

Coleman, E., Sosonkina, M., & Chow, E. (2017). Fault tolerant variants of the fine-grained parallel incomplete LU factorization. *Proceedings of the 25th High Performance Computing Symposium.*

Davis, T. (1997). University of Florida Sparse Matrix Collection. *NA Digest, 97*(23). https://www.cise.ufl.edu/research/sparse/matrices

Doa, E. H., Bhrawy, A. H., & Ezz-Eldien, S. S. (2012). A new Jacobi operational matrix: An application for solving fractional differential equations. *Applied Mathematical Modelling*, *36*(10), 4931–4943. doi:10.1016/j.apm.2011.12.031

Farina, M., & Scattolini, R. (2012). Distributed predictive control: A non-cooperative algorithm with neighbour-to-neighbor communication for linear systems. *Automatica*, *48*(6), 1088–1096. doi:10.1016/j.automatica.2012.03.020

Gutknecht, M. H. (1993). Variants of Bicgstab for Matrices with Complex Spectrum. *SIAM Journal on Scientific Computing*, *14*(5), 1020–1033. doi:10.1137/0914062

Hageman, L. A., & Young, D. M. (1981). *Applied iterative methods*. Academic Press.

Heroux, M. A., Bartlett, R. A., Howle, V. E., Hoekstra, R. J., Hu, J. J., Kolda, T. G., & Salinger, A. G. (2005). An overview of the Trilinos project. *ACM Transactions on Mathematical Software*, *31*(3), 397–423. doi:10.1145/1089014.1089021

Higham, N. J. (2009). Cholesky factorization. *Wiley Interdisciplinary Reviews: Computational Statistics*, *1*(2), 251–254. doi:10.1002/wics.18

Kearfott, R. B. (1990). *Preconditioners for the Interval Gauss-Seidel Method*. Academic Press.

Kılınç, D., Özçift, A., Bozyigit, F., Yıldırım, P., Yücalar, F., & Borandag, E. (2017). TTC-3600: A new benchmark dataset for Turkish text categorization. *Journal of Information Science*, *43*(2), 174–185. doi:10.1177/0165551515620551

Lee, W., Achar, R., & Nakhla, M. S. (2018). Dynamic GPU parallel sparse LU factorization for fast circuit simulation. *IEEE Transactions on Very Large Scale Integration (VLSI) Systems*, *26*(11), 2518–2529.

Manne, A. (1958). A linear programming model of the U. S. Petroleum Refining Industry. *Econometrica*, *26*(1), 67–106. doi:10.2307/1907384

Masood, A., Munir, E. U., Rafique, M. M., & Khan, S. U. (2015). HETS: Heterogeneous Edge and Task Scheduling Algorithm for Heterogeneous Computing Systems. *2015 IEEE 17th International Conference on High Performance Computing and Communications*, 1865-1870.

Matinfar, M., Zareamoghaddam, H., Eslami, M., & Saeidy, M. (2011). GMRES implementations and residual smoothing techniques for solving ill-posed linear systems. *Computers & Mathematics with Applications (Oxford, England)*, *63*(1), 1–13. doi:10.1016/j.camwa.2011.09.022

Mayooran, T., & Light, E. (2016). Applying the successive over-relaxation method to a real-world problems. *American Journal of Applied Mathematics and Statistics*, *4*(5), 113–117.

Motter, P. (2017). *Hardware awareness for the selection of optimal iterative linear solvers*. Academic Press.

Motter, P., Sood, K., Jessup, E., & Norris, B. (2015, November). Lighthouse: an automated solver selection tool. *Proceedings of the 3rd International Workshop on Software Engineering for High Performance Computing in Computational Science and Engineering*, 16-24.

Nakada, H., Martin, P., Wijesinghe, A., Shirai, H., Matsunaga, A., & Tominaga, H. (2018). An application of C/GMRES model predictive control to a diesel engine air path system. *IFAC-PapersOnLine*, *51*(31), 529–534. doi:10.1016/j.ifacol.2018.10.116

Ng, T. M., Farhang-Boroujeny, B., & Garg, H. K. (2001). An accelerated Gauss–Seidel method for inverse modelling. *Signal Processing*, *83*(3), 517–529. doi:10.1016/S0165-1684(02)00449-8

Nino-Ruiz, E. D., Sandu, A., & Deng, X. (2018). An ensemble kalman filter implementation based on modified Cholesky decomposition for inverse covariance matrix estimation. *SIAM Journal on Scientific Computing*, *40*(20), A867–A886. doi:10.1137/16M1097031

Özçift, A., Kılınç, D., & Bozyiğit, F. (2019). Application of grid search parameter optimized bayesian logistic regression algorithm to detect cyberbullying in Turkish microblog data. *Akademik Platform Mühendislik ve Fen Bilimleri Dergisi*, *7*(3), 355–361.

Powell, M. J. D. (1977). Restart procedures for the conjugate gradient method. *Mathematical Programming*, *12*(1), 241–254. doi:10.1007/BF01593790

Reiser, M. (1972). Large-scale numerical simulation in semiconductor device modelling. *Computer Methods in Applied Mechanics and Engineering*, *1*(1), 17–38. doi:10.1016/0045-7825(72)90019-9

Rutishauser, H. (1966). The Jacobi Method for Real Symmetric Matrices. *Numerische Mathematik*, *9*(1), 1–10. doi:10.1007/BF02165223

Saad, Y. (1994). ILUT: A dual threshold incomplete LU factorization. *Numerical Linear Algebra with Applications*, *1*(4), 387–402. doi:10.1002/nla.1680010405

Saad, Y., & Schultz, M. H. (1986). GMRES: A Generalized Minimal Residual Algorithm for Solving Nonsymmetric Linear Systems. *SIAM Journal on Scientific and Statistical Computing*, *7*(3), 856–869. doi:10.1137/0907058

Song, J., Cotilla-Sanchez, E., Ghanavati, G., & Hines, P. (2016). Dynamic Modeling of Cascading Failure in Power Systems. *IEEE Transactions on Power Systems*, *31*(3), 2085–2095. doi:10.1109/TPWRS.2015.2439237

Sood, K., Norris, B., & Jessup, E. (2017, December). Comparative performance modeling of parallel preconditioned krylov methods. In *2017 IEEE 19th International Conference on High Performance Computing and Communications; IEEE 15th International Conference on Smart City; IEEE 3rd International Conference on Data Science and Systems (HPCC/SmartCity/DSS)* (pp. 26-33). IEEE. 10.1109/HPCC-SmartCity-DSS.2017.4

Spielman, D. A. (2011). Algorithms, graph theory, and linear equations in Laplacian matrices. *Proceedings of the International Congress of Mathematicians 2010 (ICM 2010)*, 2698-2722. 10.1142/9789814324359_0164

Srinivas. (1983). Optimal Parallel Scheduling of Gaussian Elimination DAG's. *IEEE Transactions on Computers, 32*(12), 1109-1117.

Sun, Z., Tian, Y., & Wang, J. (2017). A novel projected Fletcher-Reeves conjugate gradient approach for finite-time optimal robust controller of linear constraints optimization problem: Application to bipedal walking robots. *Optimal Control Applications & Methods*, *39*(1), 130–159. doi:10.1002/oca.2339

Tucker, J. D., Miller, M. K., & Young, G. A. (2015). Assessment of thermal embrittlement in duplex stainless steels 2003 and 2205 for nuclear power applications. *Acta Materialia*, *87*, 15–24. doi:10.1016/j.actamat.2014.12.012

Virieux, J., & Operto, S. (2009). An overview of full-waveform inversion in exploration geophysics. *Geophysics*, *74*(6), WCC1–WCC26. doi:10.1190/1.3238367

Widrow, B., & Stearns, S. D. (1985). *Adaptive Signal Processing*. Prentice-Hall.

Yildirim, P., & Birant, D. (2018, September). The relative performance of deep learning and ensemble learning for textile object classification. In *2018 3rd International Conference on Computer Science and Engineering (UBMK)* (pp. 22-26). IEEE. 10.1109/UBMK.2018.8566611

Zhang, L., Cui, N., Liu, M., & Zhao, Y. (2011). Asynchronous Filtering of Discrete-Time Switched Linear Systems with Average Dwell Time. *IEEE Transactions on Circuits and Systems. I, Regular Papers*, *58*(5), 1109–1118. doi:10.1109/TCSI.2010.2092151

Zhang, Z. Q., Liu, Q. H., Xiao, C., Ward, E., Ybarra, G., & Joines, W. T. (2003). Microwave breast imaging: 3-D forward scattering simulation. *IEEE Transactions on Biomedical Engineering*, *50*(10), 1180–1189. doi:10.1109/TBME.2003.817634 PMID:14560772

Chapter 9
The Methodology of Project-Oriented Learning at the Postgraduate Level for the Training of Student Competencies

Edgar Oliver Cardoso Espinosa
https://orcid.org/0000-0001-7588-9439
Instituto Politécnico Nacional, Mexico

Jésica Alhelí Cortés Ruiz
https://orcid.org/0000-0002-5459-4874
Instituto Politécnico Nacional, Mexico

María Elena Zepeda Hurtado
Instituto Politécnico Nacional, Mexico

ABSTRACT

The objective of the chapter is the elaboration of a methodological proposal for project-oriented learning at the postgraduate level in order to develop students' mathematical competencies. The methodology of project-oriented learning has as its central element the elaboration of a project based on a problem to propose a solution through a research process. Thus, it recognizes the importance of integrating both participation and collaborative work in the training practices of graduate students to achieve quality training in their learning, which allows a transfer of skills in their work performance. In the same way, the integration of the digital competence, communication, interpersonal, and critical thinking is incorporated.

DOI: 10.4018/978-1-7998-3868-5.ch009

INTRODUCTION

The Fourth Industrial Revolution is characterized by the application of automated systems in production processes with the aim of creating digital networks that allow increasing production and using resources more efficiently. Thus, the Internet of things, connectivity and automation are incorporated. The first is aimed at improving predictive maintenance to prevent equipment failures before they occur, while the second allows the connection of billions of people to enable online commerce, social networks and mobile applications. Meanwhile, the third one is that the machinery works autonomously and flexibly (González, 2017). Therefore, the Fourth Industrial Revolution according to Schwab (2016) is described as an organization of production processes based on technology and devices that communicate with each other autonomously along the value chain.

On the other hand, Alzaga and Larreina (2016) establish that its main characteristics are: 1) Existence of a flexible model that allows the connection between suppliers, clients and employees in the productive chain of a good or service; b) Based on re-configurability that means the ability to quickly adapt to changes in customer needs; c) The digitization of production processes; d) Use of digital tools by employees in order to increase productivity. Based on the above, education has been recognized as the strategic factor that encourages the development of a society, so in the presence of the Fourth Industrial Revolution, educational systems need to consider their needs and demands, as well as the interests of citizens and the goals that the country is trying to achieve. So, teachers face the challenge of training people who are going to work in a society characterized by uncertainty, innovation and the circulation of information from the use of technological tools, which is why it is necessary for new generations to learn to live in this context (Bauman, 2008).

Therefore, it is relevant to recognize that today's society demands competent citizens who are capable of improving both their living conditions and the country in a context of dynamism, change and uncertainty (Bauman, 2003). Consequently, educational centers have to transform to respond to this new context in which learning enables the development of various capacities in the individual that allow them not only to solve the various problems of the real world but also the possibility of interacting collaboratively (Rekalde & García, 2015).

In the same way, the incorporation of the Internet into the various activities of both individuals and organizations, gave rise to the Fourth Industrial Revolution, which is transforming economies and societies by redefining the way in which people work, live and interact as a society (WEF, 2018). As Yogeshwar (2019) mentions, in the Fourth Industrial Revolution, turbo modernity is accessed, which refers to the fact that the power of the digital revolution is present in the various sectors of the country together with the generation of innovations that modify processes and work procedures.

Based on this, the Fourth Industrial Revolution offers as main opportunities: 1) Lower barriers between inventors and markets; 2) More active role for artificial intelligence; 3) Integration of different techniques and domains (fusion); 4) Quality of life and, 5) life connected through the use of the Internet (Anderson, 2012; Jules, 2017). This fusion of technologies goes beyond blending because it creates new markets and new growth opportunities for each participant in innovation. In addition, the use of the Internet will allow the interconnection of networks of physical devices, which will have an impact on an advanced connectivity of devices, systems and services with a variety of applications. So it is expected that the interconnection of these integrated devices will lead to automation in almost all areas, while generating advanced applications such as a smart grid with the ability to expand to areas such as smart cities (Gershenfeld & Vasseur, 2014).

Before this panorama, the massification of information and communication technologies (ICTs) have caused the generation of knowledge to have a very high obsolescence, so human capital should be seen as a renewable resource, which requires permanent updating and renewing with continuous learning, which will allow it to be flexible and responsive to an environment in which the only permanent thing is change (Peters, 2017).

In this sense, the OECD (2017: 1) established that "in the 21st century, competencies, skills and abilities have become the key factor to achieve individual well-being and the economic success of a society. Without adequate investment in these areas, the population languishes on the margins of society, technological progress does not transform into growth and countries are unable to compete in a global economy that is increasingly based on knowledge".

In this way, education has responded to the changing requirements of society and industry, so it represents both for institutions and therefore for teachers, the commitment to be up-to-date to enrich the comprehensive training of students and train graduates who, in addition to having a disciplinary training, are creative, innovative and proactive. From this perspective, it is necessary to understand, develop and put into practice didactic strategies in the classroom with the intention of carrying out learning experiences that strengthen and develop abilities, skills and attitudes for research, collaborative work, the use of information and communication technologies (ICTs) that allow the generation of proposals for solutions to problems in both personal and work life.

Meanwhile, UNESCO (2015) has proposed that academic training fosters a link between school content with the requirements of the labor sector with the aim of developing flexible skills necessary to the changing reality of the world so that a transfer of combined knowledge is generated together with the development of relevant capacities such as communication, digital literacy, problem solving, teamwork and an entrepreneurial spirit.

As the World Economic Forum (WEF, 2018) pointed out: the relevant competencies to develop in people for the 21st century from basic to higher education are complex problem solving, critical thinking, creativity, collaboration and digital literacy, which are essential to allow people to remain flexible and adaptable to the changing needs of society. In particular, the WEF made recommendations regarding the development of labor skills for the Fourth Industrial Revolution, which are the following: 1) Creativity; 2) People management; 3) Coordination with others; 4) Emotional intelligence; 5) Judgment and decision making; 6) Service orientation; 7) Negotiation and 8) Cognitive flexibility.

Faced with this scenario, a transformation of educational systems is necessary, where a learning environment that promotes the development of capacities as the foundation of training which approach allows the individual to solve a complex situation based on solution strategies that generate abilities, skills and attitudes is required (Pavié, 2011).

Based on the above, Pareja, Fernández and Fuentes (2019) and Rodríguez, Vargas and Luna (2010) establish that higher education institutions (HEIs) have the social function of training responsible, critical professionals who possess various skills that allow them to be competent in the workplace, which in the 21st century is characterized by the uncertainty and risks that organizations face. Thus, the success of graduates implies knowing how to make decisions in an assertive way coupled with a scenario of knowing how to work as a team to solve the various problems that arise in companies.

In this sense, HEIs have to reinvent themselves, evolve and innovate towards the construction of flexible learning experiences in the student body in relation to what, where, when and how. In addition, they will be relevant because they will allow the elimination of barriers because they will enable access to education without borders of time or space, thus increasing collaboration between people regardless of their geographic location, so generating information exchange networks (Castaño, Redecker, Vuorikari & Punie, 2014).

Specifically, higher education, and with it the postgraduate degree, has been forced to rethink itself, in order to accompany the dizzying changes that are taking place. In this way, it is expected that the training at this level is focused on the development of basic and specific competencies such as the ability to work in a team, the development of autonomous thinking and the willingness to lifelong learning (Leymonié, 2015).

In the same way, Valencia (2005) established that investment in human capital in the country generates important social benefits for the economy as a whole, which exceeds those of an individual type. These social benefits are related to the development of human capital and the development of new technologies that promote organizational innovations. In this sense, human capital comprises a diverse set of competencies that must be generated from postgraduate training in which its graduates allow the advancement of the organization, whether public or private.

To achieve this, they have begun to use active didactic methodologies in order to achieve a greater participation of students in their learning, which are oriented towards providing a protagonism of the student from doing, investigating and interacting using ICTs (Konopka, Adaime & Mosele, 2015). In this way, active methodologies are the set of activities and strategies used by teachers to achieve greater student participation in the construction of their learning, highlighting Project Oriented Learning (POL); Problem-Based Learning (PBL), Gamification and the Inverted Classroom (Roig & Álvarez, 2019).

However, specific studies on the various subjects and educational levels are required to determine the effects of POL on the academic performance of students (Landron, Agreda & Colmenero, 2018).

Hence, the objective of the chapter is the elaboration of a methodological proposal of Project Oriented Learning at the postgraduate level in order to develop the mathematical and transversal competencies of the students.

BACKGROUND

According to Van den Bergh, Mortermans, Spooren, Van Petegem, Gijbels and Vanthournout (2006), learning is a cumulative, self-regulated, directed, collaborative and individual process. At the postgraduate level, learning scientific concepts must be accompanied by various methodologies in order to produce, manage, and communicate knowledge (Rodríguez, Vargas & Luna, 2010). To achieve this, the development of a context of (re) construction of knowledge is required in which the student is provided with various opportunities to put into practice the contents to be developed in a course so that they are relevant and useful in academic training (Alptekin, Deturris & Macy, 2005).

The OECD (2012) defines competencies as the set of knowledge, abilities and skills that can be learned, allow individuals to carry out an activity or task in an adequate and systematic way, and that can be acquired and expanded through learning. This definition includes the whole range of cognitive skills such as literacy and numeracy, techniques that are specific to a sector or occupation, and the socio-emotional ones related to teamwork and communication; therefore, the set of all the skills available to the economy at any given time makes up the human capital of a country.

The mathematical competence is the set of capacities to develop and apply reasoning in order to solve situations of daily and professional life, so it is oriented towards the acquisition of using the different ways of representing information in the form of models, constructions and graphs to determine the best decision making. In this sense, it includes mastering the handling of numbers, measurements and structures to carry out the interpretation of operations and representations of a quantitative nature on personal and professional situations (Cardoso, Cortés & Cuevas, 2018).

In addition, these authors establish the importance of this competence because it allows individuals to have the ability to reason quantitatively to express in mathematical form and to use it as a series of tools for decision-making. Likewise, it includes a positive attitude towards its employment aimed at recognizing its usefulness and validity in the professional field.

To achieve the development of competencies, authors such as Dole, Bloom and Kowalske (2016), Freeman, Eddy, McDonough, Smith, Okoroafor, Jordt and Wenderoth (2014), as well as Al-Balushi and Al-Amri (2014) established the need to employ an innovative didactic approach in the development of the teaching-learning process aimed at allowing students to research from their previous knowledge in a collaborative environment. Also, in the process of training competences, the contents to be developed in the students are important, but the methodological action that is carried out is also essential, that is, the teachers' way of approaching them (Blanco, Jové & Reverter, 2012).

Merrit, Lee, Rillero and Kinach (2017) and Rosa and Martínez (2016) establish that the POL methodology offers an alternative that favors the treatment of the contents with the environment and interests of the student. For their part, Markham, Larner and Ravitz (2003) as well as Blanchard and Dolores (2016) define that POL is a systematic teaching method that involves students in the acquisition of knowledge and skills, through an extended process of inquiry, structured around questions for the elaboration of a specific product, thus articulating theory and practice and generating significant knowledge.

While Vergara (2015) conceptualizes it as a learning process based on research, the principles of egalitarian negotiation, individual responsibility, an attitude of cooperation and consensual decision-making. Meanwhile, Willard and Duffrin (2003) as well as Carrasco, Donoso, Duarte, Hernández and López (2015) established that this methodology is effective for developing skills in solving real-life situations that also involve the use of attitudes and values that promote the ability to learn throughout life.

As stated by Dole, Bloom and Kowalske (2016), POL has its foundations in the constructivist model, which considers that learning is the result of mental constructions based on previous knowledge and experiences that promote interaction between students.

As claimed by Thomas (2010) and Vergara (2015) the characteristics of POL are the following:

- It comes off situations related to the future profession of the students;
- Planning activities to develop a product; responds to the interests and needs of students;
- The result or product is valuable and useful;
- Students autonomously develop specific actions articulating theory and practice;
- The activities of the different phases of the project are decided and executed by the students themselves;
- Teamwork is encouraged so that learning is holistic and comprehensive.

Additionally, POL guidelines are as follows: Teaching occurs in the context of a situation; the key contents are learned from the context and the action, coupled with the fact that it begins with a purpose shared by the students (Doménech-Casal, 2018).

Therefore, according to Toledo and Sánchez (2018); Larmer and Mergendoller (2010) POL, for its implementation, considers the following elements:

- Public hearing: The final learning product requires a public presentation of the students by working groups using digital tools in order to address the course contents that were considered, the situation that was solved as well as the experience lived.
- Review and reflection: This strategy provides opportunities for students to make mistakes in order to continually review project completion. Thus, a follow-up process in which they request support from both the team members and the teacher is generated.
- Need of knowing: The object of study to develop the project is related to a current and relevant problem from the student's perspective, which generates a greater interest so they will be motivated to develop the solving strategy.
- Research: It involves the realization of a systematic process where a problem is raised with a specific objective, which leads to a review in various sources to collect and systematize the information to determine which one will allow to reach the solution and conclusion.

To carry out POL it is necessary for students to work autonomously in the realization of a real project during a certain period of time where they plan, design and carry out a series of activities in which it is necessary to apply various skills, attitudes, as well as effectively use available resources through reflection, debate, and communication to collect and analyze information, generating their own resolution strategies (Mettas & Constantinou, 2007). In this way, a project addresses reality so that students analyze it, use it as a learning tool, and intervene in it, so its relevance generates educational experiences that cause personal and group change (Vergara, 2016).

Hence, the benefits of POL are: it links knowing how to know with knowing how to do; enables the promotion of the skills of solution of situations, communication and autonomy; encourages lifelong learning, social responsibility and personal success; generates proactive communication and collaborative relationships; considers the needs of students with their styles, as well as motivates apathetic students (Markham, Larner & Ravitz, 2003).

In this sense, POL allows establishing the conditions that lead to active, contextualized, integrated and understanding-oriented learning, providing opportunities to reflect on the experience and apply what has been learned from a situation to be faced (Alonso, 2018); therefore, POL emphasizes the protagonism of the students in their learning process, interpersonal communication, inventiveness, the formulation and solving of situations, and the continuous search for information in order to understand their reality (Rodríguez, Vilvitskaya & Silva, 2018).

Consequently, POL is an active and student-centered methodology that is based on constructivist principles, so its main component is the elaboration of a project by carrying out various tasks through the collaboration and negotiation of the participants, who have to use research strategies (Kokotsaki, Menzies & Wiggins, 2016; García & Basilotta, 2017).

Based on the above, it is considered one of the main didactic strategies that allows the development of various competencies such as creativity, problem solving, ability to research, work collaboratively, motivation and use of social networks that are going to enhance the educational systems of the XXI century (Austin, Abella, Delgado & Hortigüela, 2016; Landron, Agreda & Colmenero, 2018). Thus, for the implementation of POL at present, an organization of spaces, materials and resources is essential, as well as the inclusion of ICTs for an effective impact on the development of student competencies (Cascales & Carrillo, 2018).

Then, this methodology allows promoting the acquisition of knowledge, oral and written communication, critical thinking, teamwork, planning and time management coupled with individual and group responsibility (De Miguel, 2006; Álvarez, Herrejón, Morelos & Rubio, 2010).

In the research carried out by Sánchez (2018), it was found that 70% of the student body positively valued teamwork based on POL, in addition to improving their academic performance. In the same way, 67% recognized the use of ICTs as an important tool to be used with POL. Similarly, 77% reported significant learning promoted by teamwork from a project. They also indicated that 95% of the product made from POL was interesting in their training coupled with the fact that 82% developed oral and written communication.

In the study by Alonso (2018) it was found that POL, from the perception of the students, allows linking the practice with the contents of a subject coupled with its recognition as useful for developing skills and identifying their future performance in the workplace. Their main findings were that 88% of the students expressed their satisfaction with the use of POL in their training, as well as 70% indicated an impact on their learning in a positive way.

In the research conducted by Alvarado, Galindo and Retamal (2018), they highlighted as their main results that 70% of the students contextualized and based their project with documents available on the internet; 100% of the participants prepared the project report; 96% developed oral and written communication, as well as 85% positively valued the use of POL in their academic training because it allowed integrating knowledge, the use of computer resources, and critical thinking.

In the study carried out by Palazuelos, San-Martín, Montoya and Fernández (2018) it was shown that POL is an active methodology for the formation of competencies from promoting greater participation, autonomous work and protagonism of the students, which is why they concluded that it is a useful proposal for the development of a set of relevant knowledge, skills, attitudes and values related to creativity, motivation and academic performance in a collaborative work environment.

In the investigation conducted by Rico, Garay and Ruiz (2018), they concluded that the implementation of POL had a positive influence on the training of students in terms of their level of knowledge and research skills, together with their analysis and synthesis capacities. It also generated teamwork that allowed them to share the challenges and achievements with their colleagues, as well as the insertion of ICTs to solve the situation, recognizing its relevance in learning.

DISCUSSION

The importance of POL is to generate greater motivation both in disciplinary issues and in the development of skills because the generation of a successful project requires that students use their communication, teamwork, time management and leadership skills (Davcev, Stojkoska, Kalajdziski & Trivodaliev, 2016).

In this sense, POL integrates with greater impact the different theoretical and practical topics, so the student implements what has been learned in the creation and development of a project, promoting intrinsic motivation, and increasing creativity (Hernando, Galán, Navarro & Rodríguez, 2011; Neebel, Merkel & Wong, 2013). Also, POL generates in the students the search, analysis and integration of the information in order to elaborate an appropriate evidence with the project so that the oral and written competence becomes a process of appropriation and transcendental handling as a discursive practice in which the way of thinking of a determined disciplinary field is disseminated and manifested (Alvarez & Boillos, 2015).

Consequently, at the postgraduate level, students have an active role in the implementation of POL, in which they are capable of generating new knowledge through the various activities they carry out, such as asking questions, investigating various sources of information, and proposal of solutions (Botella & Ramos, 2019; Boss & Krauss, 2014). In the same way, the integration in work teams of 5 to 6 members where responsibility is shared during the development of the project in which, in addition, research competencies that allow the formulation of the problem to be solved, the methodology and instruments to be used, as well as the techniques for the organization and analysis of the information obtained are promoted (Fernández, 2017; Sánchez & Vidal, 2013; Bell, Urhahne, Schanze & Ploetzner, 2010).

In addition, it allows the development of the transversal skills of planning, communication, writing and responsibility to face and solve real situations (Carrasco, Donoso, Duarte, Hernández & López, 2015; Pecore, 2012). Meanwhile, the teacher aims to create an optimal learning environment by guiding the process, encouraging the use of metacognitive strategies and guiding both individual and group tasks in order to monitor the development of the project. Besides, a process of continuous evaluation to provide both feedback and suggestions about the student's academic training should be carried out.

For the development of mathematical competencies at the postgraduate level and those of a transversal type, they are shown in the following categories organized in Table 1.

Table 1. Competencies

Mathematics	Interpersonal	Critical thinking	Communicative	Digital
Reasoning	Creativity	Documentary research	Listening	Spreadsheets
Argumentation	Negotiation	Decision making	Reading	Word processor
Interpretation	Teamwork	Leadership	Oral expression	Databases
Problem resolution			Written expression	

Source: Own elaboration

The phases of the POL methodology with their respective activities are as follows:

1st Phase: Selection and delimitation of the object of study

An assembly with group members to provide an explanation by the teacher about the didactic methodology to be used in the course is carried out. Subsequently, the characterization of the project to be built by each team considering the choice and delimitation of the topic related to the following topics: science, technology and education; sustainable development and entrepreneurship or financial management.

In a second session, the delimitation and justification is conducted in order to formulate the approach to the situation to be solved in a specific context for a given organization.

As an example, table 2 shows the planning on a mathematical content.

Table 2. Didactic planning

Thematic unit	Time series models in companies
Objective	Use time series models to predict the behavior of a company's variables
Contents	Variable identification Analysis of the components of the time series Model estimation Interpretation of results
Digital teaching resources	Research article on the concept and characteristics of a time series Video about the usefulness of time series in companies
Development	It begins with a guided discussion to characterize the time series model by the students. Selection of the administrative situation to be solved in teams Identification of the variables Research of information for the resolution of the administrative situation in teams First team exposition Research on the formulation and estimation of a time series model Second team exposition Preparation of results report in plenary
Evidence of assessment	Report and communication of results
Evaluation	Rubric on the identification of variables, estimation and interpretation of the time series model

Source: Own elaboration

2nd Phase: Search and selection of relevant information about the selected problem in journal articles and scientific books.

In this sense, to execute the documentary research it is important to consider that the characteristics to be fulfilled are its feasibility, timeliness and significance. Thus, the student body has to be guided by the teaching staff to develop the theoretical framework, collect quantitative information, as well as to organize, analyze and interpret the results to communicate them scientifically. During this process, the teacher's delivery and review of the research progress is monitored.

3rd Phase: Presentation of progress to the group

Two presentations are made in order for each team to develop the ability to communicate with the members of the group, mentioning the study variables, the type of technique and the quantitative method used for the organization and analysis of the information.

4th Phase: Elaboration of the report

The report corresponds to the main evidence that is closely related to the development of the course competencies that were mentioned in table 1. It also constitutes the support of the project prepared by each team in order to be evaluated at the end of the course.

In this sense, in the formulation of the evidence, the use of the qualities of writing is also relevant, as well as that it complies with the components of the minimum structure of an academic text: introduction, development and conclusions.

5th Phase: Presentation of the project to the group for evaluation

Each team makes the presentation and then makes a co-evaluation, as well as expressing comments on their experience of using the POL methodology to finish with their assessment and impact on academic training, identifying the skills and attitudes that were created or developed such as the organization and teamwork; responsibility; creativity; search for information and work with the use of ICTs, highlighting office tools.

Therefore, the benefits in the training of students at the postgraduate level are:

1) Link the course contents with their work performance for a transfer of learning and thus promote the integration of what is addressed in their academic life with their real life.
2) Develop soft skills such as collaboration, negotiation, decision making and time management.
3) Generate greater motivation and participation for the activities of the subject, which causes a high responsibility in its fulfillment.
4) Form mental abilities of a higher order which encourages permanent learning.
5) Promote integration and communication between members because in the labor sector it is a condition that occurs in organizations.
6) Expand the usefulness of technological tools for the development of projects.

FUTURE RESEARCH

The POL methodology has been recognized as a useful strategy to develop the diverse competencies of the student in education. Specifically, at the postgraduate level, future research is required about the degree of satisfaction provided by the experience in the POL methodology. In the same way, it is necessary to carry out future studies on other training axes that make up a postgraduate educational program, as well as the incorporation and relevance of ICTs within said methodology. Finally, it is suggested to analyze the various assessment instruments that can be used in the POL methodology in order to determine their viability and relevance in the academic training of students at the postgraduate level.

CONCLUSION

The relevance of proposing the use of the POL methodology at the postgraduate level is to integrate didactic strategies that promote not only the development of mathematical competencies but also other important ones such as creativity, negotiation and collaborative work.

Specifically, for the subject of mathematics, POL promotes learning in the student that encourages the construction of knowledge by generating motivation for being the one who makes a proposal to a real situation, for being the one who has discovered some information that was not available to him, in addition to using skills and attitudes. Thus, this methodology allows the student to recognize a quality in academic training coupled with integrating it into their work performance.

Finally, the use of POL in its implementation allows the integration of interdisciplinary learning environments and mediated by ICTs, which increase the tasks to be carried out during the project process, the information and research, which implies the contribution to motivation and communication that are available to students.

ACKNOWLEDGMENT

This research was supported by the Instituto Politécnico Nacional, México [SIP number 20200768].

REFERENCES

Al-Balushi, S., & Al-Amri, S. (2014). The effect of environmental science projects on students' environmental knowledge and science attitudes. *International Research in Geographical and Environmental Education*, *23*(3), 213–227. doi:10.1080/10382046.2014.927167

Alonso, A. (2018). Aprendizaje basado en proyectos para el desarrollo de la competencia digital docente en la formación inicial del profesorado. *Revista Latinoamericana de Tecnología Educativa, 17*(1), 9–24.

Alptekin, S., Deturris, D., Macy, D., & Ervin, J. E. (2005). Development of a flying eye: A project-based learning experience. *Journal of Manufacturing Systems, 24*(3), 226–236. doi:10.1016/S0278-6125(06)80012-9

Alvarado, H., Galindo, M., & Retamal, M. (2018). Evaluación del aprendizaje de la estadística orientada a proyectos en estudiantes de ingeniería. *Educación Matemática, 30*(3), 151–183. doi:10.24844/EM3003.07

Alvarez, M., & Boillos, M. (2015). La competencia comunicación escrita. In *Competencias genéricas en educación superior. Metodologías específicas para su desarrollo*. Narcea.

Álvarez, V., Herrejón, V., Morelos, M., & Rubio, M. (2010). Trabajo por proyectos: Aprendizaje con sentido. *Revista Iberoamericana de Educación, 52*(5), 1–13. doi:10.35362/rie5251775

Alzaga, A., & Larreina, J. (2016). *¿Qué es la industria 4.0?* Retrieved from http://www.interempresas.net/Robotica/Articulos/150900-Que-es-la-Industria-40.html

Anderson, C. (2012). *Makers: The New Industrial Revolution*. Crown Publishing.

Austin, V., Abella, V., Delgado, V., & Hortigüela, D. (2016). Aprendizaje Basado en Proyectos a través de las TIC. Una experiencia de innovación docente desde las aulas universitarias. *Formación Universitaria, 9*(3), 31–38. doi:10.4067/S0718-50062016000300005

Bauman, Z. (2003). *Modernidad líquida*. Fondo de Cultura Económica.

Bauman, Z. (2008). *Los retos de la educación en la modernidad líquida*. Gedisa.

Bell, T., Urhahne, D., Schanze, S., & Ploetzner, R. (2010). Collaborative inquiry learning: Models, tools, and challenges. *International Journal of Science Education, 3*(1), 349–377. doi:10.1080/09500690802582241

Blanchard, M., & Dolores Muzás, M. (2016). *Los proyectos de aprendizaje. Un marco metodológico para la innovación*. Narcea Ediciones.

Blanco, P., Jové, M. & Reverter, J. (2012). Paradigma estratégico para el desarrollo de habilidades competenciales Estudio descriptivo sobre la variabilidad en la percepción de habilidades competenciales de 40 alumnos de educación física en fase de formación inicial. *Educación XX1, 15*(2), 231–248.

Boss, S., & Krauss, J. (2014). *Reinventing Project-Based Learning: Your field guide to real-world projects in the digital age*. International Society for Technology in Education.

Botella, A., & Ramos, P. (2019). Investigación-acción y aprendizaje basado en proyectos. Una revisión bibligráfica. *Perfiles Educativos*, *41*(163), 127–141. http://www.scielo.org.mx/scielo.php?script=sci_arttext&pid=S0185-26982019000100127&lng=es&tlng=es

Carrasco, A., Donoso, J. A., Duarte, T., Hernández, J. J., & López, R. (2015). Diseño y validación de un cuestionario que mide la percepción de efectividad del uso de metodologías de participación activa (CEMPA). El caso del Aprendizaje Basado en Proyectos (ABP) en la docencia de la contabilidad. *Innovar (Universidad Nacional de Colombia)*, *25*(58), 125–141. doi:10.15446/innovar.v25n58.52439

Cascales, A., & Carrillo, M. (2018). Aprendizaje basado en proyectos en educación infantil: Cambio pedagógico y social. *Revista Iberoamericana de Educación*, *76*, 79–98. doi:10.35362/rie7602861

Castaño, J., Redecker, C., Vuorikari, R., & Punie, Y. (2014). Open Education 2030: Planning the future of adult learning in Europe. *Open Learning*, *28*(3), 171–186. doi:10.1080/02680513.2013.871199

Davcev, D., Stojkoska, B., Kalajdziski, S., & Trivodaliev, K. (2016). *Project Based Learning of Embedded Systems*. Retrieved from https://arxiv.org/abs/1606.07498

De Miguel, M. (2006). *Metodologías de enseñanza y aprendizaje para el desarrollo de competencias. Orientaciones para el profesorado universitario ante el Espacio Europeo de Educación Superior*. Alianza.

Dole, S., Bloom, L., & Kowalske, K. (2016). Transforming Pedagogy: Changing perspectives from teacher-centered to learner-centered. *The Interdisciplinary Journal of Problem-Based Learning*, *10*(1), 1–15. doi:10.7771/1541-5015.1538

Doménech-Casal, J. (2018). Concepciones del alumnado de secundaria sobre energía. Una experiencia de Aprendizaje Basado en Proyectos con globos aerostáticos. *Enseñanza de las Ciencias*, *36*(2), 191–213.

Fernández, M. (2017). Aprendizaje basado en proyectos en el ámbito universitario: Una experiencia de innovación metodológica en educación. *International Journal of Developmental and Educational Psychology*, *2*(1), 269–278. http://www.infad.eu/RevistaINFAD/OJS/index.php/IJODAEP/article/view/939/818

Freeman, S., Eddy, S., McDonough, M., Smith, M., Okoroafor, N., Jordt, H., & Wenderoth, M. (2014). Active learning increases student performance in science engineering and mathematics. *Proceedings of the National Acaddemy of Sciences, 111*(23), 8410–8415. doi:10.1073/pnas.1319030111 PMID:24821756

García, A., & Basilotta, A. (2017). Aprendizaje basado en proyectos (ABP): Evaluación desde la perspectiva de alumnos de educación primaria. *Revista de Investigación Educacional, 1*(35), 113–131.

Gershenfeld, N., & Vasseur, J. (2014). *As Objects Go Online: The Promise (and Pitfalls) of the Internet of Things.* Retrieved from https://www.foreignaffairs.com/articles/2014-02-12/objects-go-online

González, M. (2017). *Cuarta Revolución Industrial, empleo y bienestar social.* Real Academia de las Ciencias Morales y Políticas.

Hernando, M., Galán, R., Navarro, I., & Rodríguez, D. (2011). Ten Years of Cybertech: The Educational Benefits of Bullfighting Robotics. *IEEE Transactions on Education, 54*(4), 569–575. doi:10.1109/TE.2010.2095014

Jules, T. (2017). *Public Policy and Governance. The Global Educational Policy Environment in the Fourth Industrial Revolution: Gate, Regulated and Governed.* Emerald Group Publishing Limited.

Kokotsaki, D., Menzies, V., & Wiggins, A. (2016). Project-Based Learning: A review of the literatura. *Improving Schools, 3*(19), 267–277. doi:10.1177/1365480216659733

Konopka, C., Adaime, M., & Mosele, P. (2015). Active teaching and learning methodologies: Some considerations. *Creative Education, 6*(14), 1536–1545. doi:10.4236/ce.2015.614154

Landron, M., Agreda, M., & Colmenero, M. (2018). El efecto del aprendizaje basado en proyectos en estudiantes con altas capacidades intelectuales de una segunda lengua. *Review of Education, 380*, 210–236.

Larmer, J., & Mergendoller, J. (2010). 8 Essentials for Project-Based Learning. *Educational Leadership, 68*(1), 52–55.

Leymonié, J. (2015). Nativos e inmigrantes digitales: ¿cómo aprendemos y enseñamos? *Dixit, 12*, 10–19.

Markham, T., Larmer, J., & Ravitz, J. (2003). *Project-Based Learning Handbook: A Guide to Standards Focused Project-Based Learning for Middle and High School Teachers.* Buck Institute for Education.

Merrit, J., Lee, M., Rillero, P., & Kinach, B. (2017). Problem-Based Learning in K-8 Mathematics and Science Education: A Literature Review. *The Interdisciplinary Journal of Problem-Based Learning*, *11*(2), 1–13. doi:10.7771/1541-5015.1674

Mettas, A., & Constantinou, C. (2007). The technology fair: A project-based learning approach for enhancing problem solving skills and interest in design and technology education. *International Journal of Technology and Design Education*, *18*(1), 79–100. doi:10.100710798-006-9011-3

Neebel, D., Merkel, C., & Wong, A. (2013). Engaging a variety of students in digital design with competition. In *2013 IEEE Frontiers in Education Conference* (pp. 1091-1095). IEEE.

OECD. (2012). *Better Skills, Better Jobs, Better Lives: A Strategic Approach to Skills Policies*. OECD Publishing. doi:10.1787/9789264177338-

OECD. (2017). *Diagnóstico de la OCDE sobre la Estrategia de Competencias, Destrezas y Habilidades de México. Resumen Ejecutivo*. México: OECD. Retrieved from http://www.oecd.org/mexico/OECD-Skills-Strategy-Diagnostic-Report-Mexico.pdf

Palazuelos, E., San-Martín, P., Montoya, J., & Fernández, A. (2018). Utilidad percibida del Aprendizaje Orientado a Proyectos para la formación de competencias. *Revista de Contabilidad*, *21*(2), 150–161. doi:10.1016/j.rcsar.2017.04.004

Pareja, J., Fernández, M., & Fuentes, J. (2019). Innovación metodológica en posgrado: Aprendizaje Basado en Proyectos. *Profesorado. Revista de Currículum y Formación de Profesorado*, *23*(3), 113–128. doi:10.30827/profesorado.v23i3.9697

Pecore, J. (2012). Beyond beliefs: Teachers adapting problem-based learning to preexisting systems of practice. *The Interdisciplinary Journal of Problem-Based Learning*, *7*(2), 1–27. doi:10.7771/1541-5015.1359

Peters, M. (2017). Technological Unemployment: Educating for the Fourth Industrial Revolution. *Journal of Self-Governnace and Management Economics*, *5*(1), 25–33. doi:10.1080/00131857.2016.1177412

Rekalde, I., & García, J. (2015). El aprendizaje basado en proyectos: Un constante desafío. *Innovación Educativa (México, D.F.)*, *25*(25), 219–234. doi:10.15304/ie.25.2304

Rico, B., Garay, L., & Ruiz, E. (2018). Implementación del aprendizaje basado en proyectos como herramienta en asignaturas de ingeniería aplicada. *Revista Iberoamericana para la Investigación y el Desarrollo Educativo, 9*(17), 20–57. doi:10.23913/ride.v9i17.372

Rodríguez, E., Vargas, E., & Luna, J. (2010). Evaluación de la estrategia "Aprendizaje Basado en Proyectos". *Educación y Educadores, 13*(1), 13–25.

Rodríguez, F., Vitvitskaya, O., & Silva, M. (2018). Aprendizaje basado en proyectos en el nivel de competencias investigativas en estudiantes de Instituto Pedagógico, Trujillo, 2017. Crescendo, 9(2), 181-199.

Roig, R., & Álvarez, J. (2019). Repercusión en Twitter de las metodologías activas ABP, Flipped Classroom y Gamificación. *RIED. Revista Iberoamericana de Educación a Distancia, 22*(2), 79–96. doi:10.5944/ried.22.2.23272

Sánchez, M., & Vidal, O. (2013). Aprendizaje colaborativo basado en proyectos desarrollados en ingeniería. *Revista Iberoamericana para la Investigación y Desarrollo Educativo, 10*, 1–15.

Sánchez, N. (2018). Clase invertida y aprendizaje basado en proyectos en el aula de biología: Un proyecto de innovación para 1° de ESO. Valoración de la experiencia. *Enseñanza & Teaching, 36*(1), 81–110. doi:10.14201/et21836181110

Schwab, K. (2016). *La cuarta revolución industrial*. Penguin Ramdom House.

Thomas, J. (2010). *A review of research on Project-based learning*. The Autodesk Foundation.

Toledo, P., & Sánchez, J. (2018). Aprendizaje basado en proyectos: Una experiencia universitaria. *Profesorado. Revista de Currículum y Formación de Profesorado, 22*(2), 471–491.

UNESCO. (2015). *Replantear la educación ¿Hacia un bien común mundial?* Retrieved from https://unesdoc.unesco.org/ark:/48223/pf0000232697

Valencia, M. (2005). El capital humano, otro activo de su empresa. *Entramado, 1*(2), 20 – 33. Retrieved from https://www.redalyc.org/pdf/2654/265420471004.pdf

Van den Bergh, V., Mortermans, D., Spooren, P., Van Petegem, P., Gijbels, D., & Vanthournout, G. (2006). New assessment modes within Project-based education – the stakeholders. *Studies in Educational Evaluation, 32*(4), 345–368. doi:10.1016/j.stueduc.2006.10.005

Vergara, J. (2015). *Aprendo porque quiero. El Aprendizaje Basado en Proyectos, paso a paso.* Editorial SM: Biblioteca Innovación Educativa.

World Economic Forum. (2018b). *The Global Information Technology Report 2018.* Retrieved from http://www3.weforum.org/docs/GCR2018/05FullReport/TheGlobalCompetitivenessReport2018.pdf

World Economic Forum. (2018a). *Temas claves de la agenda global.* Retrieved from https://toplink.weforum.org/knowledge/insight/a1Gb0000000LPFfEAO/explore/summary

Yogeshwar, R. (2018). *Próxima estación futuro.* Arpa.

ADDITIONAL READING

Ausín, V., Abella, V., Delgado, V., & Hortigüela, D. (2016). Aprendizaje Basado en Proyectos a través de las TIC. Una experiencia de innovación docente desde las aulas universitarias. *Formación Universitaria, 9*(3), 31–38. doi:10.4067/S0718-50062016000300005

Domenech, J. (2017). Aprendizaje basado en proyectos y competencia científica: experiencias y propuestas para el método de estudios de caso. *Enseñanza de las ciencias: Revista de investigación y experiencias didácticas,* número extra, 5177-5184.

Domenech, J. (2019). Contexto y modelo en el aprendizaje basado en proyectos: apuntes para el ámbito científico. *Alambique: Didáctica de las ciencias experimentales,* 98, 71-76.

Fernández, M. (2017). Aprendizaje basado en proyectos en el ámbito universitario: Una experiencia de innovación metodológica en educación. *International Journal of Developmental and Educational Psychology: INFAD, 1*(2), 269–278.

Fernández, M., Polo, T., Fernández, C., Tallón, S., & Hervás, M. (2018). Voluntad de trabajo dentro del Aprendizaje basado en proyectos. *International Journal of Developmental and Educational Psychology, 1*(1), 307–316.

Fraile, J. (2017). Aprendizaje basado en proyectos y su evaluación formativa a través de rúbricas. *Revista Infancia. Educación y Aprendizaje, 3*(2), 531–535. doi:10.22370/ieya.2017.3.2.775

Lázaro, P. (2017). Innovaciones metodológicas para la sociedad digital: aprendizaje basado en proyectos, aprendizaje colaborativo, flipped classroom e inteligencias múltiples. *Tendencias pedagógicas,* 30, 339-354.

López, J. (2016). Una experiencia de aprendizaje basado en proyectos de alto contenido tecnológico. *Revista Tecnología. Ciência & Educação (Bauru), 3*, 43–62.

Osuna, I., & Rosas, A. (2017). Aprendizaje Basado en Proyectos y modelos matemáticos para estudiantes de Ingeniería en Mecatrónica. *Latin-American Journal of Physics Education, 11*(4), 1–7.

Rodríguez, V. (2017). El Aprendizaje Basado en Proyectos como herramienta para la transformación Ecosocial. *Comunicación y Pedagogía: nuevas tecnologías y recursos didácticos*, 303, 58-64.

KEY TERMS AND DEFINITIONS

Competence: Capacity that the person has in the being, knowing and knowing how to do in real situations, before a new task, to solve problems, make decisions and develop projects, from an entrepreneurial and managerial vision.

ICTs: Information and communication technologies cover all advanced technologies in manipulating and communicating information.

Internet of Things: Interconnection of everyday objects with the Internet so it uses objects equipped with embedded processors or sensors that are capable of transmitting information through the network. These connections allow remote administration, status monitoring, tracking and alerts.

Mathematical Competence: Capacity to develop and apply reasoning in order to solve situations of daily and professional life, so it is oriented towards the acquisition of using different ways of representing information in the form of models, constructions and graphs to determine the best decision making.

Project-Oriented Learning: Systematic teaching method that involves students in the acquisition of knowledge and skills, through an extended process of inquiry, structured around questions for the elaboration of a determined product for what articulates the theory and practice generating significant knowledge.

Compilation of References

Abimbade, A. (2007). *Principles and practice of Educational Technology*. International Publishers Limited.

Abrami, P. C. (2001). Understanding and promoting complex learning using technology. *Educational Research and Evaluation*, *7*(2 & 3), 113–136. doi:10.1076/edre.7.2.113.3864

Abrantes, P. (1994). *O trabalho de projecto e a relaçao dos alunos com a matemática en expreriência do projecto MAT789.* Tesis doctoral no publicada, Universidad de Lisboa, Portugal.

Adegboye, S. O. (2013). The senior secondary school science curriculum and candidates' performance. An appraisal of 1st cycle of operation. *Journal of Science Teachers Association of Nigeria. STAN*, *28*(1 & 2), 3–12.

Adelman, S., Moss, T., & Majid, A. (2005). *Data Strategy.* Prentice Hall.

Aiserman, M. A., Braverman, E. M., & Rozonoer, L. I. (1964). Theoretical foundations of the potential function method in pattern recognition. *Avtomatika i Telemehanika*, *25*, 917–936.

Ajagun, G. A. (2016). Towards good performance in science education. *Nigerian Journal of Teacher Education and Teaching*, *2*(1), 117–125.

Ajeyelemi. (2006). *New directions in the Nigerian educational system.* Being the 10th annual public lecture of the Uguogie Irowi foundation held on 8th December at NERDC conference center, Agidingbi, Ikeja, Lagos.

Akubia, L. (2011). *The effect of motivation on the performance of teachers in the Ghana Education Service: A case study of Obuasi Municipality*. KNUST.

Alarcón, J., Bonilla, E. Nava, R., Rojano, T., & Quintero, R. (1994). *Libro para el maestro. Matemáticas.* México: SEP.

Al-Balushi, S., & Al-Amri, S. (2014). The effect of environmental science projects on students' environmental knowledge and science attitudes. *International Research in Geographical and Environmental Education*, *23*(3), 213–227. doi:10.1080/10382046.2014.927167

Alonso, A. (2018). Aprendizaje basado en proyectos para el desarrollo de la competencia digital docente en la formación inicial del profesorado. *Revista Latinoamericana de Tecnología Educativa*, *17*(1), 9–24.

Alptekin, S., Deturris, D., Macy, D., & Ervin, J. E. (2005). Development of a flying eye: A project-based learning experience. *Journal of Manufacturing Systems*, *24*(3), 226–236. doi:10.1016/S0278-6125(06)80012-9

Alsina, C. (1998). *Neither a microscope nor a telescope, just a mathscope. Proceed.* International Community of Teachers of Mathematical Modelling and Applications (ICTMA).

Alterio, G., & Ruíz, C. (2010). Mediación metacognitiva, estrategias de enseñanza y procesos de pensamiento del docente de Medicina. *Educación Médica Superior*, *24*(1), 25–32.

Alvarado, H., Galindo, M., & Retamal, M. (2018). Evaluación del aprendizaje de la estadística orientada a proyectos en estudiantes de ingeniería. *Educación Matemática*, *30*(3), 151–183. doi:10.24844/EM3003.07

Alvarez, M., & Boillos, M. (2015). La competencia comunicación escrita. In *Competencias genéricas en educación superior. Metodologías específicas para su desarrollo*. Narcea.

Álvarez, V., Herrejón, V., Morelos, M., & Rubio, M. (2010). Trabajo por proyectos: Aprendizaje con sentido. *Revista Iberoamericana de Educación*, *52*(5), 1–13. doi:10.35362/rie5251775

Alzaga, A., & Larreina, J. (2016). *¿Qué es la industria 4.0?* Retrieved from http://www.interempresas.net/Robotica/Articulos/150900-Que-es-la-Industria-40.html

Anderson, C. (2012). *Makers: The New Industrial Revolution*. Crown Publishing.

Anderson, N., Chapin, S., & O'Connor, C. (2011). *Classroom discussions: Seeing math discourse in action*. Scholastic Math Solutions.

Andrei, N. (2008). Another hybrid conjugate gradient algorithm for unconstrained optimization. *Numerical Algorithms*, *47*(2), 143–156. doi:10.100711075-007-9152-9

Aragón, M. (2007). Las ciencias experimentales y la enseñanza bilingüe. *Revista Eureka sobre enseñanza y divulgación de las ciencias*, 152-175.

Aravena, M. (2001). Evaluación de proyectos para un curso de álgebra universitaria. Un estudio basado en la modelización polinómica. Tesis doctorado, Universidad de Barcelona, España.

Arslan, H. (2019). Dynamic Physarum Solver: A bio-inspired shortest path method of dynamically changing graphs. *Turkish Journal of Electrical and Computer Science*, *27*(2), 1489–1503. doi:10.3906/elk-1811-46

Arslan, H., & Manguoglu, M. (2019). A hybrid single-source shortest path algorithm. *Turkish Journal of Electrical and Computer Science*, *27*(4), 2636–2647. doi:10.3906/elk-1901-23

Arslan, H., & Manguoglu, M. (2019). A parallel bio-inspried shortest path algorithm. *Computing*, *101*(8), 969–988. doi:10.100700607-018-0621-x

Artzt, A. F., & Armour-Thomas, E. (2002). *Becoming a reflective mathematics teacher*. Lawrence Erlbaum.

Asare, J. K. (2010). *Factors Affecting Recruitment and Retention of Basic School Teachers in Obuasi Municipality*. Academic Press.

Atif, M., & Rauf, A. (2009). Efficient implementation of Gaussian Elimination method to recover generator polynomials of convolutional codes. *IEEE International Conference on Emerging Technologies (ICET 2009)*. 10.1109/ICET.2009.5353183

Aunola, K., Leskinen, E., & Nurmi, J. E. (2006). Developmental dynamics between mathematical performance, task motivation, and teachers' goals during the transition to primary school. *The British Journal of Educational Psychology, 76*(1), 21–40. doi:10.1348/000709905X51608 PMID:16573977

Austin, V., Abella, V., Delgado, V., & Hortigüela, D. (2016). Aprendizaje Basado en Proyectos a través de las TIC. Una experiencia de innovación docente desde las aulas universitarias. *Formación Universitaria, 9*(3), 31–38. doi:10.4067/S0718-50062016000300005

Awosiyan, Y. O. (2006). A survey of problems in teaching and learning of mathematics in secondary schools in Damaturu L.G.A., Yobe State. ABACUS. *The Journal of mathematics association of Nigeria, 31*(1), 130 – 141.

Aybek, B. (2001). Disiplinlerarası (bütünleştirilmiş) öğretim yaklaşımı [Interdisciplinary (integrated) teaching approach]. *Eğitim Araştırmaları Dergisi, 3*, 1–7.

Azuka, B. F. (2000). Mathematics in technological development: Focus on the next millennium – implications for secondary education. *The Journal of Mathematics Association of Nigeria., 25*(1), 74–83.

Ball, D. L. (2000). Bridging practices: Intertwining content and pedagogy in teaching and learning to teach. *Journal of Teacher Education, 51*(3), 241–247. doi:10.1177/0022487100051003013

Ball, D. L., Thames, M. H., & Phelps, G. (2008). Content knowledge for teaching: What makes it special? *Journal of Teacher Education, 59*(5), 389–407. doi:10.1177/0022487108324554

Bareiss, H. E. (1968). Sylvester's identity and multistep integer-preserving Gaussian elimination. *Mathematics of Computation, 22*, 565–578.

Basco, A., Beliz, G., Coatz, D., & Garnero, P. (2018). Industria 4.0. Fabricando el futuro, 18-19.

Batanero, C., & Díaz, C. (2010). Training teachers to teach statistics: what can we learn from research? *Statistique et enseignement, 1*(1), 5-20.

Batanero, C., Burrill, G., & Reading, C. (2011). *Teaching statistics in school mathematics—challenges for teaching and teacher education*. Springer. doi:10.1007/978-94-007-1131-0

Bauman, Z. (2003). *Modernidad líquida*. Fondo de Cultura Económica.

Bauman, Z. (2008). *Los retos de la educación en la modernidad líquida*. Gedisa.

Belbase, S. (2013). Images, anxieties, and attitudes toward mathematics. *International Journal of Education in Mathematics. Science and Technology*, *1*(4), 230–237.

Bell, T., Urhahne, D., Schanze, S., & Ploetzner, R. (2010). Collaborative inquiry learning: Models, tools, and challenges. *International Journal of Science Education*, *3*(1), 349–377. doi:10.1080/09500690802582241

Benson, D. J. (1992). Computational methods in Lagrangian and Eulerian hydrocodes. *Computer Methods in Applied Mechanics and Engineering*, *99*(2-3), 235–394. doi:10.1016/0045-7825(92)90042-I

Benzi, M. (2002). Preconditioning techniques for large systems: A Survey. *Journal of Computational Physics*, *182*(2), 418–477. doi:10.1006/jcph.2002.7176

Bergamini, M., Trifone, A., & Brozzi, G. G. (2008). Matematika verde - 1. Zanicelli.

Bhowmick, S., Eijkhout, V., Freund, Y., Fuentes, E., & Keyes, D. (2011). Application of alternating decision trees in selecting sparse linear solvers. In *Software Automatic Tuning* (pp. 153–173). Springer. doi:10.1007/978-1-4419-6935-4_10

Bishop, A. J. (1999). Mathematics teaching and values education—An intersection in need of research. *ZDM*, *31*(1), 1–4.

Bitar, A. W., Ovarlez, J., & Cheong, L. (2017). Sparsity based Cholesky factorization and its application to hyperspectral anomaly detection. *2017 IEEE 7th International Workshop on Computational Advances in Multi-Sensor Adaptive Processing (CAMSAP),* 1-5.

Bittencourt, H. R., & Clarke, R. T. (2003, July). Use of classification and regression trees (CART) to classify remotely-sensed digital images. In *IGARSS 2003. 2003 IEEE International Geoscience and Remote Sensing Symposium. Proceedings* (IEEE Cat. No. 03CH37477) (Vol. 6, pp. 3751-3753). IEEE.

Björklund, C. (2015). Pre-primary school teachers' approaches to mathematics education in Finland. *Journal of Early Childhood Education Research*, *4*(2), 69–92.

Blanchard, M., & Dolores Muzás, M. (2016). *Los proyectos de aprendizaje. Un marco metodológico para la innovación*. Narcea Ediciones.

Blanco, P., Jové, M. & Reverter, J. (2012). Paradigma estratégico para el desarrollo de habilidades competenciales Estudio descriptivo sobre la variabilidad en la percepción de habilidades competenciales de 40 alumnos de educación física en fase de formación inicial. *Educación XX1*, *15*(2), 231–248.

Blanco, R. (2013). *El pensamiento lógico desde la perspectiva de las neurociencias cognitivas* (Tesis doctoral). Universidad de Oviedo, España.

Bolognani, S., & Dörfler, F. (2015). Fast power system analysis via implicit linearization of the power flow manifold. *2015 53rd Annual Allerton Conference on Communication, Control, and Computing (Allerton)*, 402-409.

Borko, H. (2004). Professional development and teacher learning: Mapping the terrain. *Educational Researcher, 33*(8), 3–15. doi:10.3102/0013189X033008003

Boss, S., & Krauss, J. (2014). *Reinventing Project-Based Learning: Your field guide to real-world projects in the digital age*. International Society for Technology in Education.

Botella, A., & Ramos, P. (2019). Investigación-acción y aprendizaje basado en proyectos. Una revisión bibligráfica. *Perfiles Educativos, 41*(163), 127–141. http://www.scielo.org.mx/scielo.php?script=sci_arttext&pid=S0185-26982019000100127&lng=es&tlng=es

Boykin, A. W., & Noguera, P. (2011). *Creating the opportunity to learn: moving form research to practice to close the achievement gap*. ASCD.

Bozyiğit, A., Utku, S., & Nasiboğlu, E. (2019, September). Cyberbullying detection by using artificial neural network models. In *2019 4th International Conference on Computer Science and Engineering (UBMK)* (pp. 520-524). IEEE. 10.1109/UBMK.2019.8907118

Bransford, J. D., Brown, A. L., & Cocking, R. R. (2000). *How people learn: Brain, mind, experience, and school*. National Academy Press.

Braun, V., & Clarke, V. (2006). Using thematic analysis in psychology. *Qualitative Research in Psychology, 3*(2), 77–101. doi:10.1191/1478088706qp063oa

Brousseau, G. (1986). La relation didactique: le milieu. In Actes de la IVème Ecole d'Eté de Didactique des Mathématiques. IREM Paris 7.

Brown, M., & Edelson, D. (2003). *Teaching as design: Can we better understand the ways in which teachers use materials so we can better design materials to support changes in practice?* Research Report, Center for Learning Technologies in Urban Schools (Northwestern University). Retrieved from http://www.letus.org/papers.htm

Brown, J. L. (2004). *Making the most of understanding by design*. ASCD.

Brown, M. (2009). The teacher-tool relationship: Theorizing the design and use of curriculum materials. In J. T. Remillard, B. Herbel-Eisenmann, & G. Lloyd (Eds.), *Mathematics teachers at work: Connecting curriculum materials and classroom instruction* (pp. 17–36). Routledge.

Bruhn, A., Weickert, J., Feddern, C., Kohlberger, T., & Schnörr, C. (2003). Real-time optic flow computation with variational methods. In N. Petkov & M. A. Westenberg (Eds.), Lecture Notes in Computer Science: Vol. 2756. *Computer Analysis of Images and Patterns. CAIP 2003*. Springer. doi:10.1007/978-3-540-45179-2_28

Bruner, J. (2002). *Making stories: Law, literature and life*. Harvard University Press.

Bryant, D. P., Hartman, P., & Kim, S. A. (2003). Using explicit and systematic instruction to teach division skills to students with learning disabilities. *Exceptionality, 11*(3), 151–164. doi:10.1207/S15327035EX1103_03

Burns, M. (2016). *Skills for work, skills for life. I can talk 8.* Inglaterra: I CAN. Disponible en https://www.ican.org.uk/about-us/

Callahan, R. (1962). *Education and the Cult of Efficiency*. University of Chicago Press.

Callingham, R., & Watson, J. (2017). The development of statistical literacy at school. *Statistics Education Research Journal, 16*(1), 181–201. http://iase-web.org/documents/SERJ/SERJ16(1)_Callingham.pdf

Camargo-Vega, J., Camargo-Ortega, J., & Joyanes, L. (2015). Conociendo big data. *Facultad de Ingeniería, 24*(38), 63–77. doi:10.19053/01211129.3159

Cao, Y., Papageorgiou, A., Petras, I., Traub, J., & Kais, S. (2013). Quantum algorithm and circuit design solving the Poisson equation. *New Journal of Physics, 15*(1), 013021. doi:10.1088/1367-2630/15/1/013021

Carrasco, F. (2013). *Los 6 pasos que su organización debe seguir para confiar en Big Data*. América Latina. Disponible en: http://www. cioal.com/2013/07/31/los-6-pasos-que-su- organizacion-debe-seguir-para-confiar-en-big- data/

Carrasco, A., Donoso, J. A., Duarte, T., Hernández, J. J., & López, R. (2015). Diseño y validación de un cuestionario que mide la percepción de efectividad del uso de metodologías de participación activa (CEMPA). El caso del Aprendizaje Basado en Proyectos (ABP) en la docencia de la contabilidad. *Innovar (Universidad Nacional de Colombia), 25*(58), 125–141. doi:10.15446/innovar.v25n58.52439

Cascales, A., & Carrillo, M. (2018). Aprendizaje basado en proyectos en educación infantil: Cambio pedagógico y social. *Revista Iberoamericana de Educación, 76*, 79–98. doi:10.35362/rie7602861

Castaño, J., Redecker, C., Vuorikari, R., & Punie, Y. (2014). Open Education 2030: Planning the future of adult learning in Europe. *Open Learning, 28*(3), 171–186. doi:10.1080/02680513.2013.871199

Cazden, C. B. (1988). *Classroom discourse*. Heinemann.

Censor, Y. (1988). Parallel applications of block-iterative methods in medical imaging and radiation therapy. *Mathematical Programming, 42*(1-3), 307–325. doi:10.1007/BF01589408

Chuck, L. (2011). *Hadoop in Action, Publisher.* Manning Publications Co.

Chukwuneke, B. U., & Nwachkwu, C. D. (2005). State and future of education in Nigeria. In J. O. Afe & G. C. Edozie (Eds.), *Books of abstract and lead papers (pp. 58 – 59). Asaba, Nigeria: West and Solomon Publishing Company.*

Cichocki, A., & Unbehauen, R. (1992). Neural networks for solving systems of linear equations and related problems. *IEEE Transaction on Circuits and Systems- Fundamental Theory and Applications, 39*(2), 124-138.

Cinganotto, L., Guida, M., Laici, C., Mosa, E., & Panzavolta, S. (2018) MLTV, rendere l'apprendimento e il pensiero visibili nella scuola secondaria di secondo grado. *EMEM 2018 Conference Proceedings.*

Clarke, D. M., Clarke, D. J., & Sullivan, P. (2012). Reasoning in the Australian curriculum: Understanding its meaning and using the relevant language. *Australian Primary Mathematics Classroom, 17*(3), 28–32.

Cloudera, Inc. (2013). Disponible en: http://www.cloudera.com/content/cloudera/en/ why-cloudera/ hadoop-and-big-data.html

Cobb, P. (1994). Where is the mind? Constructivist and sociocultural perspectives on mathematical development. *Educational Researcher, 23*(7), 13–20. doi:10.3102/0013189X023007013

Cobb, P., & Steffe, L. P. (1983). The constructivist researcher as teacher and model builder. *Journal for Research in Mathematics Education, 14*(2), 83–94. doi:10.2307/748576

Cobb, P., & Yackel, E. (1996). Constructivist, emergent, and sociocultural perspectives in the context of developmental research. *Educational Psychologist, 31*(3), 175–190. doi:10.120715326985ep3103&4_3

Cockcroft, W. H. (1982). *Mathematics counts. Report of the Committee of Inquiry into the Teaching of Mathematics in Schools.* Her Majesty's Stationery Office.

Cohen, S. (1972). *Folk devils and moral panic.* Routledge.

Cole, M., & Cole, S. (2001). *The development of children* (4th ed.). Scientific American Books. Distributed by W.N. Freeman and Company.

Coleman, E., Sosonkina, M., & Chow, E. (2017). Fault tolerant variants of the fine-grained parallel incomplete LU factorization. *Proceedings of the 25th High Performance Computing Symposium.*

computerworld.es. (2013). Disponible en: http://www. computerworld.es/sociedad-de-la-informacion/ el-mercado-del-big-data-crecera-hasta-los- 32400-millones-de-dolares-en-2017, 2013.

Cook, K. L., & Bush, S. B. (2018). Design thinking in integrated STEAM learning: Surveying the landscape and exploring exemplars in elementary grades. *School Science and Mathematics, 118*(3–4), 93–103. doi:10.1111sm.12268

Coombes, C. *What is Effective Teaching of Mathematics?* Generation Ready. Retrieved online at https://www.generationready.com/what-is-effective-teaching-of-mathematics/

Creswell, J. W. (2012). *Educational research: Planning, conducting and evaluating quantitative and qualitative research* (4th ed.). Pearson Publication.

Cross, C., Woods, T., & Schweingruber, H. (2009). *Mathematics Learning in Early Childhood. Committee on Early Childhood Mathematics.* The National Academic Press.

Cross, N. (2001). Designerly ways of knowing: Design disipline versus design science. *Design Issues, 17*(3), 49–55. doi:10.1162/074793601750357196

Darling-Hammond, L., Chung Wei, R., Andree, A., Richardson, N., & Orphanos, S. (2009). *Professional learning in learning profession: A status report on teacher development in the United States and Abroad.* Stateford University/ National Staff Development Council.

Darling-Hammond, L., & McLaughlin, M. W. (1995). Policies that support professional development in an era of reform. *Phi Delta Kappan, 76*, 597–604.

Davcev, D., Stojkoska, B., Kalajdziski, S., & Trivodaliev, K. (2016). *Project Based Learning of Embedded Systems.* Retrieved from https://arxiv.org/abs/1606.07498

Davis, T. (1997). University of Florida Sparse Matrix Collection. *NA Digest, 97*(23). https://www.cise.ufl.edu/research/sparse/matrices

Davis, E. A. (2004). Knowledge integration in science teaching: Analyzing teachers' knowledge development. *Research in Science Education, 34*(1), 21–53. doi:10.1023/B:RISE.0000021034.01508.b8

Davis, E., & Krajcik, S. J. (2005). Designing educative curriculum materials to promote teacher learning. *Educational Researcher, 34*(3), 3–14. doi:10.3102/0013189X034003003

De Lange, J. (1995). Assessment: No change without problems. *Reform in school mathematics and authentic assessment*, 87-172.

De Miguel, M. (2006). *Metodologías de enseñanza y aprendizaje para el desarrollo de competencias. Orientaciones para el profesorado universitario ante el Espacio Europeo de Educación Superior.* Alianza.

Dejean, D. (2018). *Maths. Enseignement obligatoire et de spécialité (ES). Enseignement de spécialité (L).* Hachette Livre, Vanves Cedex.

DeKetelaere, A., & Kelchtermans, G. (1996). Collaborative curriculum development: An encounter of different professional knowledge systems. *Teachers and Teaching, 2*(1), 71–85. doi:10.1080/1354060960020106

Del Val, J. (2016). Industria 4.0: la transformación digital de la industria. *Conferencia de Directores y Decanos de Ingeniería Informática.* Retrieved from http://coddii.org/wp-content/uploads/2016/10/Informe-CODDII-Industria-4.0.pdf

Delgado, P. (2019). *Educación STEM: ¿qué es y cómo sacarle provecho? Observatorio de innovación educativa: tecnológico de Monterrey.* Retrieved from https://observatorio.tec.mx/edu-news/educacion-stem-que-es-y-como-sacarle-provecho

Desgagné, S. (1997). Le concept de recherche collaborative: l'idée d'un rapprochement entre chercheurs universitaires et praticiens enseignant. *Revue des sciences de l'éducation.*

Dewey, J. (1933). *How we think.* Heath.

Dienes, Z. P. (2015). *Building up mathematics*. Hutchinson Education Ltd.

Dijcks, J. (2013). *Oracle: Big Data for the Enterprise*. Disponible en: http://www.oracle.com/technetwork/database/bigdata-appliance/ overview/wp-bigdatawithoracle-1453236.pdf?ssSourceSiteId=ocomes

Doa, E. H., Bhrawy, A. H., & Ezz-Eldien, S. S. (2012). A new Jacobi operational matrix: An application for solving fractional differential equations. *Applied Mathematical Modelling*, *36*(10), 4931–4943. doi:10.1016/j.apm.2011.12.031

Dole, S., Bloom, L., & Kowalske, K. (2016). Transforming Pedagogy: Changing perspectives from teacher-centered to learner-centered. *The Interdisciplinary Journal of Problem-Based Learning*, *10*(1), 1–15. doi:10.7771/1541-5015.1538

Domènech, J., Lope, S. & Mora, L. (2019). Qué proyectos STEM diseña y qué dificultades expresa el profesorado de secundaria sobre Aprendizaje Basado en Proyectos. *Revista Eureka sobre Enseñanza y Divulgación de las Ciencias, 16*(2), 35-46.

Doménech-Casal, J. (2018). Concepciones del alumnado de secundaria sobre energía. Una experiencia de Aprendizaje Basado en Proyectos con globos aerostáticos. *Enseñanza de las Ciencias*, *36*(2), 191–213.

Doverborg, E., & Pramling Samuelsson, I. (2009). Grundläggande matematik. In S. Sheri-dan, I. Pramling Samuelsson, & E. Johansson (Eds.), *Barns tidiga lärande. En tvärsnittsstudie om förskolan som miljö för barns lärande* [Children's Early Learning. A Cross-sectional Study of Preschool as an Environment for Children's Learning] (pp. 125–150). Acta Universitatis Gothoburgensis.

Dubet, F. (1994). *Sociologie de l'expérience*. Le Seuil.

Dunne, D., & Roger, M. (2006). Design thinking and how it will change management education: An interview and discussion. *Academy of Management Learning & Education*, *5*(4), 512–523. doi:10.5465/amle.2006.23473212

Edelson, D. C. (2002) Design research: What we learn when we engage in design. *Journal of the Learning Sciences*, *11*(1), 105–121. doi:10.1207/S15327809JLS1101_4

Farina, M., & Scattolini, R. (2012). Distributed predictive control: A non-cooperative algorithm with neighbour-to-neighbor communication for linear systems. *Automatica*, *48*(6), 1088–1096. doi:10.1016/j.automatica.2012.03.020

Farmer, J. D., Gerretson, H., & Lassak, M. (2003). What teachers take from professional development: Cases and implications. *Journal of Mathematics Teacher Education*, *6*(4), 331–360. doi:10.1023/A:1026318709074

Fernández, M. (2017). Aprendizaje basado en proyectos en el ámbito universitario: Una experiencia de innovación metodológica en educación. *International Journal of Developmental and Educational Psychology*, *2*(1), 269–278. http://www.infad.eu/RevistaINFAD/OJS/index.php/IJODAEP/article/view/939/818

Flavell, J. H. (1979). Metacognition and cognitive monitoring: A new area of cognitive developmental inquiry. *The American Psychologist*, *34*(10), 906–911. doi:10.1037/0003-066X.34.10.906

Foro Económico Mundial sobre América Latina América Latina en un punto de inflexión: construyendo una nueva narrative. (2018). Available from http://www3.weforum.org/docs/LA18_Report_Spanish.pdf

Freeman, S., Eddy, S., McDonough, M., Smith, M., Okoroafor, N., Jordt, H., & Wenderoth, M. (2014). Active learning increases student performance in science engineering and mathematics. *Proceedings of the National Acaddemy of Sciences*, *111*(23), 8410–8415. doi:10.1073/pnas.1319030111 PMID:24821756

Fuller, F. F. (1969). Concerns of teachers: A developmental conceptualization. *American Educational Research Journal*, *6*(2), 207–226. doi:10.3102/00028312006002207

Funke, J. (2010). Complex problema solving: A case for complex cognition? *Cognitive Processing*, *11*(2), 133–142. doi:10.100710339-009-0345-0 PMID:19902283

Gal, I. (2002). Adults' statistical literacy: Meanings, components, responsibilities. *International Statistical Review*, *70*(1), 1–25. doi:10.1111/j.1751-5823.2002.tb00336.x

García, A., & Basilotta, A. (2017). Aprendizaje basado en proyectos (ABP): Evaluación desde la perspectiva de alumnos de educación primaria. *Revista de Investigación Educacional*, *1*(35), 113–131.

García, M., & García, V. (2020). Metodología STEAM y su uso en Matemáticas para estudiantes de bachillerato en tiempos de pandemia Covid-19. *Revista Científica (Maracaibo)*, *6*(2), 163–180.

Gardner, H. (1983). *Frames of Mind: The Theory of Multiple Intelligences*. Basic Books.

Garet, M. S., Porter, A. C., Desimone, L., Birman, B. F., & Yoon, K. S. (2001). What makes professional development effective? Lessons from a national sample of teachers. *American Educational Research Journal*, *38*(4), 915–945. doi:10.3102/00028312038004915

Garfield, J. (1995). How Students Learn Statistics. *International Statistical Review*, *63*(1), 25–34. doi:10.2307/1403775

Garfield, J., & Ahlgren, A. (1988). Difficulties in Learning Basic Concepts in Probability and Statistics: Implications for Research. *Journal for Research in Mathematics Education*, *19*(1), 44–63. doi:10.2307/749110

Gay, L. R. (2005). *Educational Research. Competencies for Analysis and Application* (5th ed.). Merrill Publication Company.

George, P. S. (2005). A rationale for differentiated instruction in the regular classroom. *Theory into Practice*, *44*(3), 185–193. doi:10.120715430421tip4403_2

Gershenfeld, N., & Vasseur, J. (2014). *As Objects Go Online: The Promise (and Pitfalls) of the Internet of Things*. Retrieved from https://www.foreignaffairs.com/articles/2014-02-12/objects-go-online

Ginsburg, H. (2002). Little children, big mathematics: Learning and teaching in the pre-school. In *Proceedings of the 26th Conference of the International Group for the Psychology of Mathematics Education* (Vol. 1, pp. 3–14). University of East Anglia.

Gitonga, J. M. (2000). *A study of the problems that teacher trainers in primary teachers colleges in the Eastern Province of Kenya face in preparing Trainees to teach Mathematics* (Published M.Ed. thesis). University of Nairobi Press.

Glynn, S., & Koballa, T. R. (2005). *The contextual teaching and learning instructional approach* (R. E. Yager, Ed.). National Science Teachers Association Press.

Goldman, S., Carroll, M. P., & Kabayadondo, Z. (2012). Assessing d.Learning: Capturing the journey of becoming a design thinker. In H. Plattner, C. Meinel, & L. Leifer (Eds.), Design thinking research. Measuring performance in context (pp. 13–33). Berlin: Springer.

Goldman, S., & Zielezinski, M. B. (2016). Teaching with design thinking: Developing new vision and approaches to twenty-first century learning. In L. A. Annetta & J. Minogue (Eds.), *Connecting science and engineering education practices in meaningful ways* (pp. 237–262). Springer. doi:10.1007/978-3-319-16399-4_10

Gómez, J. (2003). La Modelización Matemática. Una herramienta válida en la enseñanza de las Matemáticas Universitarias. *Sumas*, *42*, 37–35.

González, M. (2017). *Cuarta Revolución Industrial, empleo y bienestar social*. Real Academia de las Ciencias Morales y Políticas.

Gouvea, J. S., Sawtelle, V., Geller, B., & Turpen, C. (2013). A Framework for Analyzing Interdisciplinary Tasks: Implications for Student Learning and Curricular Design. *CBE Life Sciences Education*, *12*(2), 197–205. doi:10.1187/cbe.12-08-0135 PMID:23737627

Gross-Tsur, V., Manor, O., & Shalev, R. S. (2006). Developmental dyscalculia: Prevalence and demographic features. *Developmental Medicine and Child Neurology*, *37*, 906–914.

Guida, M. (2018). Rendere visibili il pensiero degli studenti ed il loro apprendimento in matematica. In La didattica della matematica, strumento concreto in aula. Atti del XXXII Convegno Nazionale Incontri con la matematica Castel San Pietro (Bo).

Guida, M. (2019). La formazione del docente di matematica. *Essere a Scuola*, *6*(201), 14-19.

Guida, M., & Cinganotto, L. (2020). Visible Teacher Thinking and Group Learning. In M. Rehm, O. Mealha, & T. Rebedea (Eds.), *Ludic, Co-design and Tools Supporting Smart Learning Ecosystems and Smart Education*. Springer Nature.

Gutknecht, M. H. (1993). Variants of Bicgstab for Matrices with Complex Spectrum. *SIAM Journal on Scientific Computing*, *14*(5), 1020–1033. doi:10.1137/0914062

Hadoop.apache.org. (2013). Disponible en: http:// hadoop.apache.org/

Hafiz, T. J., & Hina, H. A. (2016). Causes of poor performance in mathematics from teachers, parents and student's perspective. = *American Scientific Research Journal for Engineering, Technology, and Sciences*, *15*(1), 122–136.

Hageman, L. A., & Young, D. M. (1981). *Applied iterative methods*. Academic Press.

Hannon, V. (2017). *Thrive: Schools Reinvented for the Real Challenges We Face*. Innovation Unit Press.

Haure, J.-P., & Lavandier, F. (2018). *Annales Bac, 2018. Sujets et Corrigés*. Hachette Livre, Vanves Cedex.

Hernando, M., Galán, R., Navarro, I., & Rodríguez, D. (2011). Ten Years of Cybertech: The Educational Benefits of Bullfighting Robotics. *IEEE Transactions on Education*, *54*(4), 569–575. doi:10.1109/TE.2010.2095014

Heroux, M. A., Bartlett, R. A., Howle, V. E., Hoekstra, R. J., Hu, J. J., Kolda, T. G., & Salinger, A. G. (2005). An overview of the Trilinos project. *ACM Transactions on Mathematical Software*, *31*(3), 397–423. doi:10.1145/1089014.1089021

Hiebert, J., Carpenter, T., Fennema, K., Fuson, D., & Wearne, D. (1997). Making Sense: teaching and learning mathematics with understanding. Portsmouth, NH: Heinemann.

Higham, N. J. (2009). Cholesky factorization. *Wiley Interdisciplinary Reviews: Computational Statistics*, *1*(2), 251–254. doi:10.1002/wics.18

Hoover, J., & Patton, J. (2004). Differentiating Standards-Based Education for students with Diverse Needs. *Remedial and Special Education*, *25*(2), 74–78. doi:10.1177/07419325040250020101

Hopkins, B. (2011). *Beyond the Hype of Big Data*. Disponible en: http://www.cio.com/article/692724/Beyond_the_Hype_of_Big_Data.

Hoyles, C. (2002). Mathematical skills in the workplace: Final report to the Science Technology and Mathematics Council. Inglaterra: Institute of Education, University of London.

Hughes, K. L., Bailey, T. R., & Mechur, M. J. (2001). *School-to-Work: Making a difference in education. A Research Report to America*. Institute on Education and the Economy.

Huizinga, T., Handelzalt, A., Nieveen, N., & Voogt, J. (2015). Fostering teachers' design expertise in teacher design teams: Conducive design and support activities. *Curriculum Journal*, *26*(1), 137–163. doi:10.1080/09585176.2014.990395

IBM. (2012). *Big Data and analytics platform*. Disponible en: http://www-01.ibm. com/software/data/bigdata

Instituto Politécnico Nacional. (2004). *Materiales para Reforma. Un Nuevo Modelo educativo para el IPN*. IPN.

Instituto Politécnico Nacional. (2008). *Programa de estudios de la Unidad de Aprendizaje de Probabilidad y estadística*. IPN.

IPN. (2006). Geometría Analítica. Libro del Profesor. México: IPN.

Jablonka, E. (2003). Mathematical Literacy. In Second International Handbook of Mathematics Education. Springer. doi:10.1007/978-94-010-0273-8_4

John, O. N. (2018). WAEC gives reasons there is mass failure in 2018 May/June WASSCE result. *Daily Post*. Retrieved on February 11, 2020 at https://dailypost.ng/2018/07/17/waec-gives-reasons-mass-failure-2018-may-june-wassce-result/

Johnson, E. (2002). *Contextual teaching and learning*. Corwin Press.

Joyanes, L. (2013). *Big Data: Análisis de grandes volúmenes de datos en organizaciones*. Editorial Alfaomega.

Jules, T. (2017). *Public Policy and Governance. The Global Educational Policy Environment in the Fourth Industrial Revolution: Gate, Regulated and Governed*. Emerald Group Publishing Limited.

Kakai, B. M. (2011). *Challenges facing students in adapting to mathematics learning in secondary schools in Bungoma South District, Kenya* (Unpublished master's thesis). Kenatta University.

Kearfott, R. B. (1990). *Preconditioners for the Interval Gauss-Seidel Method*. Academic Press.

Kılınç, D., Özçift, A., Bozyigit, F., Yıldırım, P., Yücalar, F., & Borandag, E. (2017). TTC-3600: A new benchmark dataset for Turkish text categorization. *Journal of Information Science*, *43*(2), 174–185. doi:10.1177/0165551515620551

Kokotsaki, D., Menzies, V., & Wiggins, A. (2016). Project-Based Learning: A review of the literatura. *Improving Schools*, *3*(19), 267–277. doi:10.1177/1365480216659733

Konopka, C., Adaime, M., & Mosele, P. (2015). Active teaching and learning methodologies: Some considerations. *Creative Education*, *6*(14), 1536–1545. doi:10.4236/ce.2015.614154

Kothari, C. R. (2004). *Research methodology: Methods and techniques*. New Age International Limited Publishers.

Krechevsky, M., Mardell, B., Rivard, M., & Wilson, D. (2013). *Visible Learners: Promoting Reggio-Inspired Approaches in All Schools*. John Wiley & Sons.

Landron, M., Agreda, M., & Colmenero, M. (2018). El efecto del aprendizaje basado en proyectos en estudiantes con altas capacidades intelectuales de una segunda lengua. *Review of Education*, *380*, 210–236.

Land, S. M., & Hannafin, M. J. (2006). Student-centered learning environments. In D. H. Jonassen & S. M. Land (Eds.), *Theoretical foundations of learning environments* (pp. 1–23). Lawrence Erlbaum Associates, Inc.

Larmer, J., & Mergendoller, J. (2010). 8 Essentials for Project-Based Learning. *Educational Leadership*, *68*(1), 52–55.

Lee, W., Achar, R., & Nakhla, M. S. (2018). Dynamic GPU parallel sparse LU factorization for fast circuit simulation. *IEEE Transactions on Very Large Scale Integration (VLSI) Systems*, *26*(11), 2518–2529.

Lempert-Shepell, E. (1995). Teacher self-identification in culture from Vygotsky's developmental perspective. *Anthropology & Education Quarterly*, *26*(4), 425–442. doi:10.1525/aeq.1995.26.4.05x1062v

Lerkkanen, M. K., Kiuru, N., Pakarinen, E., Viljaranta, J., Poikkeus, A. M., Rasku-Puttonen, H., Siekkinen, M., & Nurmi, J. E. (2012). The role of teaching practices in the development of children's interest in reading and mathematics in kindergarten. *Contemporary Educational Psychology*, *37*(4), 266–279. doi:10.1016/j.cedpsych.2011.03.004

Lesh, R., & Lehrer, R. (2003). Models and Modeling Perspective on the Development of Students and Teachers. *Mathematical Thinking and Learning*, *5*(2-3), 2–3, 109–129. doi:10.1080/10986065.2003.9679996

Levine, M. (2002). *A mind at a time*. Simon & Schuster.

Lewis, S., & Batts, K. (2005). How to implement differentiated instruction? Adjust, adjust, adjust. *Journal of Staff Development*, *26*(4), 26–31.

Leymonié, J. (2015). Nativos e inmigrantes digitales: ¿cómo aprendemos y enseñamos? *Dixit*, *12*, 10–19.

Lin, L., Shadiev, R., Hwang, W. Y., & Shen, S. (2020). From knowledge and skills to digital works: An application of design thinking in the information technology course. *Thinking Skills and Creativity*, *36*, 36. doi:10.1016/j.tsc.2020.100646

Little, C., & Jones, K. (2007). Contexts for pure mathematics: An analysis of A level mathematics papers. *Proceedings of the British Society for Research into Learning Mathematics*, *27*(1), 48–53.

Little, M. E. (2009). Teaching mathematics: Issues and solutions. *Teaching Exceptional Children Plus*, *6*(1), 12–22.

Luka, I. (2014). Design thinking in pedagogy. *Journal of Education Culture and Society*, *2*, 63–74.

Maccini, P., Mulcahy, C. A., & Wilson, M. G. (2007). A follow-up of mathematics interventions for secondary students with learning disabilities. *Learning Disabilities Research & Practice*, *22*(1), 58–74. doi:10.1111/j.1540-5826.2007.00231.x

Mackay, C. B. (2001). *Republic of Kenya: Second University in Kenya, Report of Presidential working Party*. Government Printers.

Magnoler, P. (2012). *Ricerca e formazione. La professionalizzazione degli insegnanti*. PensaMultimedia.

Manne, A. (1958). A linear programming model of the U. S. Petroleum Refining Industry. *Econometrica*, *26*(1), 67–106. doi:10.2307/1907384

Markham, T., Larmer, J., & Ravitz, J. (2003). *Project-Based Learning Handbook: A Guide to Standards Focused Project-Based Learning for Middle and High School Teachers*. Buck Institute for Education.

Marzano, R. J., Pickering, D. J., & Pollock, J. E. (2001). *Classroom instruction that works: Research based strategies for increasing student achievement*. Association for Supervision and Curriculum Development.

Masood, A., Munir, E. U., Rafique, M. M., & Khan, S. U. (2015). HETS: Heterogeneous Edge and Task Scheduling Algorithm for Heterogeneous Computing Systems. *2015 IEEE 17th International Conference on High Performance Computing and Communications*, 1865-1870.

Matinfar, M., Zareamoghaddam, H., Eslami, M., & Saeidy, M. (2011). GMRES implementations and residual smoothing techniques for solving ill-posed linear systems. *Computers & Mathematics with Applications (Oxford, England)*, *63*(1), 1–13. doi:10.1016/j.camwa.2011.09.022

Mau, S. T., & Harkness, S. S. (2020). The role of teacher educators and university supervisors to help student teachers reflect: From monological reflection toward dialogical conversation. *Reflective Practice*, *21*(2), 1–12. doi:10.1080/14623943.2020.1716710

Mayer-Schönberger, V., & Cukier, K. (2013). *Big data: la revolución de los datos masivos*. Turner.

Mayooran, T., & Light, E. (2016). Applying the successive over-relaxation method to a real-world problems. *American Journal of Applied Mathematics and Statistics*, *4*(5), 113–117.

McTighe, J., & O'Connor, K. (2006, Summer). Seven practices for effective learning. *Educational Leadership*, *63*(10), 13–19.

McTighe, J., & Thomas, R. S. (2003). Backward design for forward action. *Educational Leadership*, *60*(5), 52–55.

McTighe, J., & Wiggins, G. (2004). *Understanding by design: Professional development workbook*. ASCD.

Medrich, E., Calderon, S., & Hoachlander, G. (2002). *Contextual teaching and learning strategies in high schools: Developing a vision for support and evaluation*. Mpr Associates, Incorporated.

Mercer, N., & Littleton, K. (2007). *Dialogue and the development of children's thinking: a sociocultural approach*. Routledge. doi:10.4324/9780203946657

Merlot Pedagogy. (2013). *A program of the California State University in partnership with higher education institutions*. Author.

Merrit, J., Lee, M., Rillero, P., & Kinach, B. (2017). Problem-Based Learning in K-8 Mathematics and Science Education: A Literature Review. *The Interdisciplinary Journal of Problem-Based Learning*, *11*(2), 1–13. doi:10.7771/1541-5015.1674

Metin, Y. (2015). High school students' attitudes towards mathematics. *Eurasia Journal of Mathematics, Science & Technology Education, 12*(4), 931-945.

Mettas, A., & Constantinou, C. (2007). The technology fair: A project-based learning approach for enhancing problem solving skills and interest in design and technology education. *International Journal of Technology and Design Education, 18*(1), 79–100. doi:10.100710798-006-9011-3

Meyer, C. L. (2006). *Learning to teach conceptually: Four preservice teachers' journeys* (Unpublished doctoral dissertation). University at Albany.

Mezirow, J. (Ed.). (2000). The Jossey-Bass Higher and Adult Education Series *Learning as Transformation: Critical Perspectives on a Theory in Progress*. Jossey- Boss.

Miheso, M. M. (2002). *Factors affecting mathematics performance among Secondary Schools Students in Nairobi Province, Kenya* (Unpublished M.Ed. thesis). Kenyatta University.

Miller, S., & Mercer, C. (2017). The educational aspects of mathematical disabilities. *Journal of Learning Disabilities, 30*(1), 47–56. doi:10.1177/002221949703000104 PMID:9009876

Ministry of Education and Science of Bulgaria. (2016a). *School Educational Programme in Mathematics for the 2nd grade*. https://www.mon.bg/bg/1997

Ministry of Education and Science of Bulgaria. (2016b). *School Educational Programme in Mathematics for the 6th grade*. https://www.mon.bg/bg/1998

Ministry of Education and Science of Bulgaria. (2016c). *School Educational Programme in Mathematics for the 7th grade*. https://www.mon.bg/bg/1690

Ministry of Education and Science of Bulgaria. (2016d). *Educational Programmes in Mathematics by School grades*. https://www.mon.bg/bg/28

Ministry of Education and Science of Bulgaria. (2017). *School Educational Programme in Mathematics for the 3rd grade*. https://www.mon.bg/bg/1689

Mislina, A., & Fauziah, K. (2017). Assessing Mathematics' Attitude among Technical University Students. *International Journal of Academic Research in Business & Social Sciences, 7*(4), 103–115.

Montoro, S. (2012). *Server and Cloud Platform*. Disponible en: http://lapastillaroja.net/2012/02/nosql-for-non-programmers/

Montoya, S. (2018). *Defining Literacy. GAML Fifth Meeting*. UNESCO Institute of Statistics. http://gaml.uis.unesco.org/wp-content/uploads/sites/2/2018/12/4.6.1_07_4.6-defining-literacy.pdf

Moore, D. S. (1997). New Pedagogy and New Content: The Case of Statistics. *International Statistical Review, 65*(2), 123–165. doi:10.1111/j.1751-5823.1997.tb00390.x

Motter, P. (2017). *Hardware awareness for the selection of optimal iterative linear solvers*. Academic Press.

Motter, P., Sood, K., Jessup, E., & Norris, B. (2015, November). Lighthouse: an automated solver selection tool. *Proceedings of the 3rd International Workshop on Software Engineering for High Performance Computing in Computational Science and Engineering*, 16-24.

Muller, E., & Burkhardt, H. (2007). Applications and modelling for mathematics—Overview. In Modelling and Applications in Mathematics Education (pp. 267-274). Springer.

Murray, T., & Arroyo, I. (2002). *Towards measuring and maintaining the zone of proximal development*. Plenum Press.

Muvango, W. M. (2013). *Teachers have to be more creative*. Saturday Nation.

Mzomwe, Y. M., Calkin, S. M., & Respickius, O. C. (2019). Investigating Students' Attitude towards Learning Mathematics. *International Electronic Journal of Mathematics Education, 14*(1), 207–231.

Nakada, H., Martin, P., Wijesinghe, A., Shirai, H., Matsunaga, A., & Tominaga, H. (2018). An application of C/GMRES model predictive control to a diesel engine air path system. *IFAC-PapersOnLine, 51*(31), 529–534. doi:10.1016/j.ifacol.2018.10.116

National Center for Education Statistics. (2004). *Highlights from the trends in International mathematics and science study (TIMSS)*. U.S. Government Printing Office.

National Research Council. (2001). *Adding It Up: Helping Children Learn Mathematics*. Washington, DC: The National Academies Press. doi:10.17226/9822

National Soft Skills Association. (2016). *Skills employers seek*. Disponible en https://www.nationalsoftskills.org/

Nations United. (2015). *Asamblea General*. Retrieved from https://www.un.org/sustainabledevelopment/es/

Neebel, D., Merkel, C., & Wong, A. (2013). Engaging a variety of students in digital design with competition. In *2013 IEEE Frontiers in Education Conference* (pp. 1091-1095). IEEE.

Nelson, D., & Annetta, L. A. (2016). Creating Disruptive Innovators: Serious Educational Game Design on the Technology and Engineering Spectrum. In Connecting Science and Engineering Education Practices in Meaningful Ways (pp. 3-17). Springer.

Newman, M. A. (1977). An analysis of sixth-grade pupils' errors on written mathematical tasks. In M. A. Clements & J. Foyster (Eds.), Research in Mathematics education in Australia (Vol. 2). Melbourne: Swinburne College Press.

Newton Public Schools. https://www.newton.k12.ma.us/

Ng, T. M., Farhang-Boroujeny, B., & Garg, H. K. (2001). An accelerated Gauss–Seidel method for inverse modelling. *Signal Processing, 83*(3), 517–529. doi:10.1016/S0165-1684(02)00449-8

Nieveen, N. (2009). Formative evaluation in educational design research. In T. Plomp & N. Nieveen (Eds.), *An Introduction to Educational Design Research* (pp. 89–102). SLO.

Nino-Ruiz, E. D., Sandu, A., & Deng, X. (2018). An ensemble kalman filter implementation based on modified Cholesky decomposition for inverse covariance matrix estimation. *SIAM Journal on Scientific Computing*, *40*(20), A867–A886. doi:10.1137/16M1097031

Niss, M. (1989). Aim and of applications and modeling in mathematics curricula. In *Applications and modeling in learning and teaching mathematics* (pp. 22–32). Ellis Horwood.

Noble, C. L. (2011). *How does understanding by design influence student achievement in eight grade social studies?* (Unpublished doctoral dissertation). Capella University.

Nordlund, M. (2003). *Differentiated instruction: Meeting the educational needs of all students in your classroom*. Scarecrow Education.

Nworgu, B. G. (2007). Reforms and quality education in Nigeria. A key note address. In *19th annual conference of the association for promoting quality education in Nigeria*. River State University of Education, Port Harcourt.

Obomanu, B. J., & Adaramola, M. O. (2011). Factors related to underachievement in Mathematics, Science, and Technology education (MSTE) in secondary schools in Rivers State, Nigeria. *World Journal of Education*, *1*(1), 102–105. doi:10.5430/wje.v1n1p102

OECD. (2006). *PISA Science Competencies for Tomorrow's World*. OECD iLibrary. https://www.oecd-ilibrary.org/education/pisa_19963777

OECD. (2009). *PISA 2009 Assessment Framework. Key competencies in reading, Mathematics and science*. Retrieved May 9, 2020, from https://www.oecd.org/pisa/pisaproducts/44455820.pdf

OECD. (2012). *Better Skills, Better Jobs, Better Lives: A Strategic Approach to Skills Policies*. OECD Publishing. doi:10.1787/9789264177338-

OECD. (2017). *Diagnóstico de la OCDE sobre la Estrategia de Competencias, Destrezas y Habilidades de México. Resumen Ejecutivo*. México: OECD. Retrieved from http://www.oecd.org/mexico/OECD-Skills-Strategy-Diagnostic-Report-Mexico.pdf

OECD. (2018). *PISA 2015 Results in Focus*. https://www.oecd.org/pisa/pisa-2015-results-in-focus.pdf

Okebukola, P. (2002). *Beyond the stereotype to new trajectories in science teaching*. Special lecture presented at the 43rd annual conference of STAN and CAMSTE.

Olavsrud, T. (2012). *Big Data Causes Concern and Big Confusion*. Disponible en:http:// www.cio.com/article/700804/Big_Data_Causes_Concern_and_Big_Confusion?page=2&taxonomy Id=3002

Olubusiyi, P. (2003). *Education and manpower. Poor teachers' pay, welfare rebound in students' performance*. Vangard Newspaper.

Opara, I. J., Magnus-Arewa, E. A., & Nwaukwu, C. (2017). Predictors of students' attitude towards mathematics in Obioakpor local government area of rivers state, Nigeria. *International Journal of Education, Learning and Development, 5*(7), 73–85.

Osuna, Mariela, Ruiz, Amparo, Ortega, Alberto, & Salas. (2019). Retos de ingeniería: Enfoque educativo STEM+A en la revolución industrial 4.0. *Innovación Educativa (México, D.F.), 19*(80), 15–32.

Oyeniran, J. O. (2013). Teaching methods. An introduction to principles and methods of teaching. Lagos: SIBIS Ventures.

Özçift, A., Kılınç, D., & Bozyiğit, F. (2019). Application of grid search parameter optimized bayesian logistic regression algorithm to detect cyberbullying in Turkish microblog data. *Akademik Platform Mühendislik ve Fen Bilimleri Dergisi, 7*(3), 355–361.

Özdemir, E., & Üzel, D. (2011). Gerçekçi matematik eğitiminin öğrenci başarısına etkisi ve öğretime yönelik öğrenci görüşleri [The effect of realistic mathematics education on student achievement and student views on teaching]. *Hacettepe University Journal of Education, 40*, 332–343.

Palazuelos, E., San-Martín, P., Montoya, J., & Fernández, A. (2018). Utilidad percibida del Aprendizaje Orientado a Proyectos para la formación de competencias. *Revista de Contabilidad, 21*(2), 150–161. doi:10.1016/j.rcsar.2017.04.004

Pande, M., & Bharathi, V. (2020). Theoretical foundations of design thinking – A constructivism learning approach to design thinking. *Thinking Skills and Creativity, 36*, 36. doi:10.1016/j.tsc.2020.100637

Panzavolta, S., Mosa, E., Laici, C., Guida, M., & Cinganotto, L. (2018). MLTV, Rendere L'apprendimento e il Pensiero Visibili Nella Scuola Secondaria di Secondo Grado. *Exploring the Micro, Meso and Macro navigating between dimensions in the digital learning landscape EDEN 2018 Annual Conference Proceedings.*

Pareja, J., Fernández, M., & Fuentes, J. (2019). Innovación metodológica en posgrado: Aprendizaje Basado en Proyectos. *Profesorado. Revista de Currículum y Formación de Profesorado, 23*(3), 113–128. doi:10.30827/profesorado.v23i3.9697

Pecore, J. (2012). Beyond beliefs: Teachers adapting problem-based learning to preexisting systems of practice. *The Interdisciplinary Journal of Problem-Based Learning, 7*(2), 1–27. doi:10.7771/1541-5015.1359

Peter, J., Martin, W., David, H., & Daphne, C. (2017). The sustainable development goals and Information and communication technologies. *Indonesian Journal of Sustainability Accounting and Management, 1*(1), 1–15. doi:10.28992/ijsam.v1i1.22

Peters, M. (2017). Technological Unemployment: Educating for the Fourth Industrial Revolution. *Journal of Self-Governnace and Management Economics, 5*(1), 25–33. doi:10.1080/00131857.2016.1177412

Piaget, J. (1952). *The child's concept of number*. Humanities Press.

Pimm, D. (1987). *Speaking mathematically*. Routledge.

Plugge, E., Membrey, P., & Hawkins, T. (2010). *The Definitive Guide to MongoDB: The NoSQL Database for Cloud and Desktop Computing*. Published Apress Media LLC.

Pollard, A., Collins, J., Simco, N., Swaffield, S., Warin, J., & Warwick, P. (2005). *Reflective teaching*. Continuum.

Postholm, M. B. (2008). Teachers developing practice: Reflection as key activity. *Teaching and Teacher Education*, *24*(7), 17–28. doi:10.1016/j.tate.2008.02.024

Powell, M. J. D. (1977). Restart procedures for the conjugate gradient method. *Mathematical Programming*, *12*(1), 241–254. doi:10.1007/BF01593790

Powell, W., & Napoliello, S. (2005). Using observation to improve instruction. *Educational Leadership*, *62*(5), 52–55.

Project Zero's Thinking Routine Toolbox. (n.d.). Retrieved May 9, 2020, from https://pz.harvard.edu/thinking-routines

Putman, R., & Borko, H. (2000, January - February). What do new views of knowledge and thinking have to say. *Educational Researcher*, *29*(1), 4–15. doi:10.3102/0013189X029001004

Redmond, E., & Wilson, J. (2012). *Seven Databases in Seven Weeks*. O'Reilly Media, Inc., Pragmatic Programmers, LLC.

Reiser, M. (1972). Large-scale numerical simulation in semiconductor device modelling. *Computer Methods in Applied Mechanics and Engineering*, *1*(1), 17–38. doi:10.1016/0045-7825(72)90019-9

Rekalde, I., & García, J. (2015). El aprendizaje basado en proyectos: Un constante desafío. *Innovación Educativa (México, D.F.)*, *25*(25), 219–234. doi:10.15304/ie.25.2304

Remillard, J. T. (2005). Examining key concepts in research on teachers' use of mathematics curricula. *Review of Educational Research*, *75*(2), 211–246. doi:10.3102/00346543075002211

Rico, B., Garay, L., & Ruiz, E. (2018). Implementación del aprendizaje basado en proyectos como herramienta en asignaturas de ingeniería aplicada. *Revista Iberoamericana para la Investigación y el Desarrollo Educativo*, *9*(17), 20–57. doi:10.23913/ride.v9i17.372

Ritchhart, R. (2015). *Creating cultures of thinking*. Jossey-Bass.

Ritchhart, R., Church, M., & Morrison, K. (2011). *Making Thinking Visible*. Jossey Bass.

Rodríguez, F., Vitvitskaya, O., & Silva, M. (2018). Aprendizaje basado en proyectos en el nivel de competencias investigativas en estudiantes de Instituto Pedagógico, Trujillo, 2017. Crescendo, 9(2), 181-199.

Rodríguez, E., Vargas, E., & Luna, J. (2010). Evaluación de la estrategia "Aprendizaje Basado en Proyectos". *Educación y Educadores*, *13*(1), 13–25.

Roig, R., & Álvarez, J. (2019). Repercusión en Twitter de las metodologías activas ABP, Flipped Classroom y Gamificación. *RIED. Revista Iberoamericana de Educación a Distancia, 22*(2), 79–96. doi:10.5944/ried.22.2.23272

Russom, P. (2012). *Big Data Analytics, TDWI*. The Data Warehousing Institute.

Rutishauser, H. (1966). The Jacobi Method for Real Symmetric Matrices. *Numerische Mathematik, 9*(1), 1–10. doi:10.1007/BF02165223

Ryan, K., & Bohlin, K. E. (1999). *Building character in schools*. Jossey Bass.

Saad, Y. (1994). ILUT: A dual threshold incomplete LU factorization. *Numerical Linear Algebra with Applications, 1*(4), 387–402. doi:10.1002/nla.1680010405

Saad, Y., & Schultz, M. H. (1986). GMRES: A Generalized Minimal Residual Algorithm for Solving Nonsymmetric Linear Systems. *SIAM Journal on Scientific and Statistical Computing, 7*(3), 856–869. doi:10.1137/0907058

Saenz, C. (2009). The role of contextual, conceptual and procedural knowledge in activating mathematical competencies (PISA). *Educational Studies in Mathematics, 71*(2), 123–143. doi:10.100710649-008-9167-8

Salgado, M. (2014). *Oracle apuesta por Big Data con tecnología y proyectos*. Disponible en: https://www.computerworld.es/big-data/oracle-apuesta-por-big-data-con-tecnologia-y-proyectos

Sam, L., & Ernest, P. (1997). Values in mathematics education: What is planned and what is espoused? *British Society for Research into Learning Mathematics. Proceedings of the Day Conference held at University of Nottingham*, 37-44.

Sánchez, M., & Vidal, O. (2013). Aprendizaje colaborativo basado en proyectos desarrollados en ingeniería. *Revista Iberoamericana para la Investigación y Desarrollo Educativo, 10*, 1–15.

Sánchez, N. (2018). Clase invertida y aprendizaje basado en proyectos en el aula de biología: Un proyecto de innovación para 1° de ESO. Valoración de la experiencia. *Enseñanza & Teaching, 36*(1), 81–110. doi:10.14201/et2183618110

Saunders, M., Lewis, P., & Thornhill, A. (2007). *Research methods for business students* (4th ed.). Prentice Hall.

Schield, M. (2011). Statistical literacy: A new mission for data producers. *Statistical Journal of the IAOS*, (27), 173–183. https://www.researchgate.net/publication/228968797

Schleicher, A. (2019). *PISA 2019: Insights and Interpretation*. OECD.

Schön, D. A. (1993). *Il professionista riflessivo: per una nuova epistemologia della pratica professionale*. Edizioni Dedalo.

Schornick, P. (2010). Looking at High School Mathematics Education From the Inside Out. *National Association of Secondary School Principals (NASSP) Bulletin., 94*(1), 17–39.

Schramn, W. (1977). *Big media, little media, tools and techniques for instructions*. Sage Publications.

Schussler, D. L., Stooksberry, L. M., & Bercaw, L. A. (2010). Understanding teacher candidate dispositions: Reflecting to build self-awareness. *Journal of Teacher Education*, *61*(4), 350–363. doi:10.1177/0022487110371377

Schwab, K. (2016). *La cuarta revolución industrial*. Penguin Ramdom House.

Sharma, P. (2019). Digital Revolution of Education 4.0. International Journal of Engineering and Advanced Technology. *International Journal of Engineering Research & Technology (Ahmedabad)*, *9*(2).

Shulman, L. S. (1986). Those who understand: Knowledge growth in teaching. *Educational Researcher*, *15*(2), 4–14. doi:10.3102/0013189X015002004

Shulman, L. S. (1987). Knowledge and teaching: Foundations of the new reform. *Harvard Educational Review*, *57*(1), 1–22. doi:10.17763/haer.57.1.j463w79r56455411

Simmie, G. M. (2007). Teacher design teams (TDTs) building capacity for innovation, learning and curriculum implementation in the continuing professional development of in-career teachers. *Irish Educational Studies*, *26*(2), 163–176. doi:10.1080/03323310701295914

Simon, M. A. (2000). Research on mathematics teacher development: The teacher development experiment. In R. Lesh & E. Kelly (Eds.), *Research design in mathematics and science education* (pp. 335–359). Lawrence Erlbaum.

Smorti, A. (1994). *Il pensiero narrativo. Costruzione di storie e sviluppo della coscienza sociale*. Giunti.

Song, J., Cotilla-Sanchez, E., Ghanavati, G., & Hines, P. (2016). Dynamic Modeling of Cascading Failure in Power Systems. *IEEE Transactions on Power Systems*, *31*(3), 2085–2095. doi:10.1109/TPWRS.2015.2439237

Sood, K., Norris, B., & Jessup, E. (2017, December). Comparative performance modeling of parallel preconditioned krylov methods. In *2017 IEEE 19th International Conference on High Performance Computing and Communications; IEEE 15th International Conference on Smart City; IEEE 3rd International Conference on Data Science and Systems (HPCC/SmartCity/DSS)* (pp. 26-33). IEEE. 10.1109/HPCC-SmartCity-DSS.2017.4

Sowder, J. T. (2007). The mathematical education and development of teachers. In F. K. Lester Jr., (Ed.), *Second handbook of research on mathematics teaching and learning: A project of the national council of teachers of mathematics* (pp. 157–223). Information Age.

Spielman, D. A. (2011). Algorithms, graph theory, and linear equations in Laplacian matrices. *Proceedings of the International Congress of Mathematicians 2010 (ICM 2010)*, 2698-2722. 10.1142/9789814324359_0164

Srinivas. (1983). Optimal Parallel Scheduling of Gaussian Elimination DAG's. *IEEE Transactions on Computers, 32*(12), 1109-1117.

Steel, C. F. (2009). *The inspired teacher: How to know one, grow one, or to be one*. ASCD.

Stodolsky, S. S. (2008). *The subject matters*. University of Chicago Press.

Suarez, D. (2007). Differentiation by challenge: Using a tiered program of instruction in mathematics. In W. Powel & O. Kusuma (Eds.), *Making the difference: differentitation in international schools* (pp. 199–227). ASCD.

Sugareva M., Mourgova, M. (2017, June). *Учебният материал по вероятности и статистика в учебните програми в средните училища в България* [The teaching material in Probabilities and statistics in the curricula of secondary education in Bulgaria] [Paper presentation]. Conference of the Union of Bulgarian Mathematicians "Pedagogical and Technological Aspects of Educational Computer Games", Plovdiv, Bulgaria

Sugareva, M., & Mourgova, M. (2018, April). Образованието по статистика в училищата в България – предпоставка за статистическа грамотност сред населението [School Education in Statistics in Bulgaria – a prerequisite for statistical literacy of the population]. In *Proceedings of the Forty-seventh Spring Conference of the Union of Bulgarian Mathematicians* (pp. 259 – 265). http://www.math.bas.bg/smb/2018_PK/tom_2018/index.html

Sullivan, P., Zevenbergen, Z., & Mousley, J. (2003). The contexts of mathematics tasks and the context of the classroom: Are we including all students? *Mathematics Education Research Journal, 15*(2), 107–121. doi:10.1007/BF03217373

Sun, Z., Tian, Y., & Wang, J. (2017). A novel projected Fletcher-Reeves conjugate gradient approach for finite-time optimal robust controller of linear constraints optimization problem: Application to bipedal walking robots. *Optimal Control Applications & Methods, 39*(1), 130–159. doi:10.1002/oca.2339

Supovitz, J. A., & Turner, H. M. (2000). The effects of professional development on science teaching practices and classroom culture. *Journal of Research in Science Teaching, 37*(9), 963–980. doi:10.1002/1098-2736(200011)37:9<963::AID-TEA6>3.0.CO;2-0

Sztajn, P., Wilson, H., Edgington, C., & Myers, M. (2013). Using design experiments to conduct research on mathematics professional development. *Alexandria Revista de Educação em Ciência e Tecnologia, 6*(1), 9–34.

Tabachnick, B., & Zeichner, K. (Eds.). (1991). Issues and practices in inquiry oriented teacher education. New York: The Falmer Press.

Tata, U. S., Abba, A., & Abdullahi M. S. (2014). The causes of poor performance in mathematics among public senior secondary school students in Azare Metropolis of Bauchi State, Nigeria. *IOSR Journal of Research & Method in Education, 4*(6), 32-40.

Taylor, D. (1996). *The Healing Power of Stories*. Doubleday.

The Report of the Expert Panel on Mathematics in Grades 4 to 6 in Ontario. (2004). *Teaching and Learning Mathematics*. Author.

Thomas, J. (2010). *A review of research on Project-based learning*. The Autodesk Foundation.

Thompson, A. G. (1984). The relationship of teachers' conceptions of mathematics and mathematics teaching to instructional practice. *Educational Studies in Mathematics*, *15*(2), 105–127. doi:10.1007/BF00305892

Thousand, J. S., Villa, R. A., & Nevin, A. I. (2007). *Differentiating instruction: Collaborative planning and teaching for universally designed learning*. Corwin Press.

Tieso, C. (2005). The effects of grouping practices and curricular adjustments on achievement. *Journal for the Education of the Gifted*, *29*(1), 60–89. doi:10.1177/016235320502900104

Tieso, C. L. (2003). Ability grouping is not just tracking anymore. *Roeper Review*, *26*(1), 29–36. doi:10.1080/02783190309554236

Tishkovskaya, S., & Lancaste, G. A. (2012). Statistical Education in the 21st Century: A Review of Challenges, Teaching Innovations and Strategies for Reform. *Journal of Statistics Education: An International Journal on the Teaching and Learning of Statistics*, *20*(2). Advance online publication. doi:10.1080/10691898.2012.11889641

Toffler, A. (1970). *Future shock*. Bantam Books.

Toker, Z. (2016). *A preservice mathematics teacher's reflective practices on self-improvement regarding teaching and learning process in practice teaching* (Unpublished doctoral dissertation). METU, Turkey.

Toledo, P., & Sánchez, J. (2018). Aprendizaje basado en proyectos: Una experiencia universitaria. *Profesorado. Revista de Currículum y Formación de Profesorado*, *22*(2), 471–491.

Tomlinson, C. A. (1999). *The Differentiated Classroom: Responding to the Needs of all Learners*. ASCD.

Tomlinson, C. A., & Kalbfleisch, M. L. (1998). Teach me, teach my brain: A call for differentiated classrooms. *Educational Leadership*, *56*(3), 52–55.

Tomlinson, C. A., & McTighe, J. (2006). *Integrating differentiated instruction and understanding by design*. ASCD.

Trigwell, K., & Prosser, M. (1996). Changing approaches to teaching: A relational perspective. *Studies in Higher Education*, *21*(3), 275–284. doi:10.1080/03075079612331381211

Trilling, B., & Fadel, C. (2009). *21st century skills: Learning for life in our times*. John Wiley & Sons.

Tsakiridou, H., & Vavyla, E. (2019). Probability Concepts in Primary School. *American Journal of Educational Research*, *3*(4), 535–540. doi:10.12691/education-3-4-21

Tucker, J. D., Miller, M. K., & Young, G. A. (2015). Assessment of thermal embrittlement in duplex stainless steels 2003 and 2205 for nuclear power applications. *Acta Materialia*, *87*, 15–24. doi:10.1016/j.actamat.2014.12.012

Turkish Education Association (TED). (2009). *Öğretmen yeterlikleri* [Teacher Competencies]. Türk Eğitim Derneği.

UKCES. (2011). *The supply of and demand for high-level STEM skills*. Briefing Paper. Commission for Employment and Skills.

Ulusoy, K., & Dilmaç, B. (2016). *Değerler eğitimi* [Value education]. Pegem Yayıncılık.

UNESCO. (2011). *ICT Competency Framework for Teachers*. París: UNESCO. Disponible en http://unesdoc.unesco.org/images/ 0021/002134/213475e.pdf

UNESCO. (2015). *Replantear la educación ¿Hacia un bien común mundial?* Retrieved from https://unesdoc.unesco.org/ark:/48223/pf0000232697

UNESCO. (2017) *Coloquio Internacional y Foro sobre Políticas de la UNESCO. Descifrar el código: la educación de las niñas en materia de Ciencias, Tecnología, Ingeniería y Matemáticas*. Retrieved from https://es.unesco.org/coloquio-internacional-y-foro-politicas-unesco-descifrar-codigo-educacion-ninas-materia-ciencias

UNESCO. (2020) *STEM Resources*. Retrieved from http://www.unesco.org/new/en/natural-sciences/science-technology/engineering/engineering-education/stem-resources/

Valencia, M. (2005). El capital humano, otro activo de su empresa. *Entramado*, *1*(2), 20 – 33. Retrieved from https://www.redalyc.org/pdf/2654/265420471004.pdf

Van den Bergh, V., Mortermans, D., Spooren, P., Van Petegem, P., Gijbels, D., & Vanthournout, G. (2006). New assessment modes within Project-based education – the stakeholders. *Studies in Educational Evaluation*, *32*(4), 345–368. doi:10.1016/j.stueduc.2006.10.005

Vappula, H., & Clausen-May, T. (2006). Context in maths test questions: Does it make a difference? *Research in Mathematics Education*, *8*(1), 99–115. doi:10.1080/14794800008520161

Vergara, J. (2015). *Aprendo porque quiero. El Aprendizaje Basado en Proyectos, paso a paso*. Editorial SM: Biblioteca Innovación Educativa.

Virieux, J., & Operto, S. (2009). An overview of full-waveform inversion in exploration geophysics. *Geophysics*, *74*(6), WCC1–WCC26. doi:10.1190/1.3238367

Vos, P., & Kuiper, W. (2005). Trends (1995-2000) in the TIMSS mathematics performance assessment in the Netherlands. *Educational Research and Evaluation*, *11*(2), 141–154. doi:10.1080/13803610500110794

Vygotsky, L. S. (1962). *Thought and Language*. MIT Press. doi:10.1037/11193-000

Vygotsky, L. S. (1978). *Mind in society: The Development of Higher Psychological Processes* (M. Cole, V. John-Steiner, S. Scribner, & E. Souberman, Eds.). Harvard University Press.

Vygotsky, L. S. (1978). *Mind in society: The development of higher psychological processes*. Harvard University Press.

Warford, M. K. (2011). The zone of proximal teacher education. *Teaching and Teacher Education*, *27*(2), 252–258. doi:10.1016/j.tate.2010.08.008

Watanabe, R., & Ischinger, B. (2009). *Learning Mathematics for Life: A Perspective from PISA*. OECD Publishing.

Watson, J., & Callingham, R. (2003). Statistical literacy: A complex hierarchical construct. *Statistics Education Research Journal*, *2*(2), 3–46.

Wells, G. (2000). *Dialogic inquiry: Towards a socio-cultural practice and theory of education*. Cambridge University Press.

Wendy, E. A. (2018). *Drop in WASSCE performance worrying*. Retrieved on February 11, 2020 at https://www.myjoyonline.com/news/2019/March-12th/drop-in-wassce-performance-worrying.php

Widrow, B., & Stearns, S. D. (1985). *Adaptive Signal Processing*. Prentice-Hall.

Wiggins, G., & McTighe, J. (2005). Understanding by Design (2nd ed.). Alexandria, VA: ASCD.

Wongsopawiro, D. S. (2012). *Examining science teachers' pedagogical content knowledge in the context of a professional development program* (Unpublished doctoral dissertation). Leiden: ICLON, Leiden University Graduate School of Teaching.

World Bank. (2018). *World development report 2018: Learning to realize education's promise*. International Bank for Reconstruction and Development, The World Bank.

World Economic Forum. (2018a). *Temas claves de la agenda global*. Retrieved from https://toplink.weforum.org/knowledge/insight/a1Gb0000000LPFfEAO/explore/summary

World Economic Forum. (2018b). *The Global Information Technology Report 2018*. Retrieved from http://www3.weforum.org/docs/GCR2018/05FullReport/TheGlobalCompetitivenessReport2018.pdf

Wubbels, T., Korthagen, F., & Broekman, H. (1997). Preparing teachers for realistic mathematics education. *Educational Studies in Mathematics*, *32*(1), 1–28. doi:10.1023/A:1002900522457

Yildirim, P., & Birant, D. (2018, September). The relative performance of deep learning and ensemble learning for textile object classification. In *2018 3rd International Conference on Computer Science and Engineering (UBMK)* (pp. 22-26). IEEE. 10.1109/UBMK.2018.8566611

Yıldırım, A. (1996). Disiplinlerarası öğretim kavramı ve programlar açısından doğurduğu sonuçlar [Interdisciplinary teaching concept and its consequences in terms of programs]. *Hacettepe Üniversitesi Eğitim Fakültesi Dergisi*, *12*, 89–94.

Yogeshwar, R. (2018). *Próxima estación futuro*. Arpa.

Yusuf, S., & Araba, H. (2019). Perceived causes of students' failure in mathematics in Kwara state junior secondary schools: Implication for educational managers. *International Journal of Educational Studies in Mathematics*, *6*(1), 19–33.

Zakaria, E., & Iksan, Z. (2007). Promoting cooperative learning in science and mathematics education: A Malaysian perspective. *Eurasia Journal of Mathematics, Science and Technology Education*, *3*(1), 35–39. doi:10.12973/ejmste/75372

Zambrano, K. (2017). Fortalecimiento de las matemáticas a través de las STEAM en la Tecnoacademia de Neiva. *Revista Ciencias Humanas*, *14*, 39–52.

Zan, R. (2017). Il ruolo cruciale del pensiero narrativo nella comprensione dei problemi. *Didattica della matematica. Dalla ricerca alle pratiche d'aula, 2*, 46-57. doi:10.33683/ddm.17.2.3

Zanzali, A. (2003). Implementing the intended mathematics curriculum: Teachers' beliefs about the meaning and relevance of problem solving. *Proceedings of the International Conference: The Decidable and the Undecidable in Mathematics Education.*

Zdnet.com. (2010). *Big Data.* Disponible en: http:// www.zdnet.com/search?q=big+data

Zee, E. V., & Minstrell, J. (1977). Using questioning to guide students thinking. *Journal of the Learning Sciences*, *6*(2), 227–269.

Zeichner, K., & Liston, D. (1996). *Reflective teaching: an introduction.* Lawrence Erlbaum Associates.

Zhang, L., Cui, N., Liu, M., & Zhao, Y. (2011). Asynchronous Filtering of Discrete-Time Switched Linear Systems with Average Dwell Time. *IEEE Transactions on Circuits and Systems. I, Regular Papers*, *58*(5), 1109–1118. doi:10.1109/TCSI.2010.2092151

Zhang, Z. Q., Liu, Q. H., Xiao, C., Ward, E., Ybarra, G., & Joines, W. T. (2003). Microwave breast imaging: 3-D forward scattering simulation. *IEEE Transactions on Biomedical Engineering*, *50*(10), 1180–1189. doi:10.1109/TBME.2003.817634 PMID:14560772

Zikopoulos, P., Eaton, C., DeRoos, D., Deutsch, T., & Lapis, G. (2012). *Understanding Big Data.* McGraw-Hill Books.

Zikopoulos, P., deRoos, D., Parasuraman, K., Deutsch, T., Corrigan, D., & Giles, J. (2013). *Harness the Power of Big Data.* McGraw-Hill Companies.

Zingiswa, J. (2019). *Mathematics Education System in South Africa.* Retrieved at https://www.intechopen.com/books/education-systems-around-the-world/mathematics-education-system-in-south-africa

About the Contributors

Edgar Oliver Cardoso Espinosa, PhD, is Professor - Researcher since 2009 on Division of Postgraduate Studies and Research of the Superior School of Commerce and Administration (ESCA) of the National Polytechnic Institute (IPN), Mexico City. In his position conducts research projects focused on the management and development of education in educational programs at senior level. He teaches courses Methodological Seminar I, II, III and Statistics for Educational Research to Master's and Research Seminars I, II and III as well as Research Visits I, II, III and IV to PhD level. Dr. Cardoso earned his doctorate at the IPN. He has a master of science in management and development of education in the ESCA-IPN. His research includes evaluation of educational programs and executive education. He belongs to the National System of Researchers with the level I distinction in the CONACyT of Mexico. He has directed several research projects in the area of educational management. He has conducted thesis addresses at the master's and doctoral level in the educational area.

* * *

Salako E. Adekunle received a degree (B.Eng) in Electrical and Computer Engineering, Master of Technology (M.Tech) in Computer Science from the Federal University of Technology, Minna, Niger State. He is a professional member of the Nigeria Computer Society and Teachers Registration Council of Nigeria (TRCN). His research interests include Biometric Security, Educational Technology and Control Technology. He had attended a number of local and international conferences and seminars. He has published papers in reputable national, international peer-reviewed journals and conference proceedings. He has authored four academic textbooks: Introduction to Computer Logic, Learning Pascal Made Easy, A Handbook on Symbolic Logic and BASIC Programming Language, co-authored a computer textbook. He is a member of Educational Services on Institution-Based Research, Academic Manuscript/Books Development (AMBD) and Publication of Academic/Research Journals.

Alma Alicia Benítez Pérez has a doctorate in science with a specialty. Researcher SNI I. Professor of the Instituto Politécnico Nacional.

Betsabé Adalia Contreras Domínguez is a mathematics teacher at CECYT No. 11 "Wilfrido Massieu" in the Instituto Politécnico Nacional.

Jésica Alhelí Cortés Ruíz, PhD, is Professor - Researcher since 2018 on la Sección de Estudios de Posgrado e Investigación of the Escuela Superior de Comercio y Administración (ESCA) of the Instituto Politécnico Nacional (IPN), Mexico City. PhD in Administrative Sciences from the Higher School of Commerce and Administration, Master of Science in Systems Engineering from the Higher School of Mechanical and Electrical Engineering. She is a specialist in transdisciplinary systemic and has been the author of chapters of dissemination books, as well as various articles in research journals. Her systemic training and dedication to the knowledge society has allowed her to teach In the International Business Degree, the Master in Administration in Management and Development of Education, The Master in Business Administration for Sustainability And the Master of Science in Business Administration.

Maria Guida is a researcher at INDIRE, a public research institute whose main mission is to sustain the evolution of the Italian educational system through technology-enhanced teachers training, system actions for improvement and innovation. At INDIRE Maria has had an active role in teacher training National Programs (i. e. the mandatory training programs for Newly Qualified Teachers). She collaborated with the University of Florence too, as a tutor for Mathematics and Physics postgraduate students in the TFA, the specialization course for soon-to-be teachers. She has been working for years (2007 onward) in the research "Immersive Education", which investigates 3D virtual environments and their educational potential. At European level she is involved in STEM and Mathematics Education, with a focus on the active and student-centered learning, which she has driven for years her research to, because she could identify it, through studies and experiences, as an element of strong contrast to the dropout phenomenon.

Marta Sugareva is Doctor of Sociological sciences since 2000. Research fellow in the Institute of Demography (now - Institute for Population and Human Studies) at the Bulgarian Academy of sciences since 1980. Teaching courses of "Statistics", " Demography", and "Family Sociology" in Sofia university "St. Kliment Ohridski" (Department of Sociology) since 1980 up to 2009. Since 2009 - full professor of "Statistics and Demography" in Plovdiv university "Paisii Hilendarski". Author of numerous scientific publications in Bulgarian, English and French, on various subjects.

Zerrin Toker completed her undergraduate, graduate, and doctorate education at Middle East Technical University, in Elementary Education. She worked as a Mathematics Teacher between 2004-2017 and as the Head of the Department of Mathematics in 2010-2017. As a Professional Development Coordinator, she assumed in her institution between 2011-2013, worked as a program for math programs from kindergarten to 8th grade, as well as in-service training and development of teachers. She has over 10 years of teaching experience in the International Baccalaureate Primary Years Program and the Middle Years Program. In addition, she wrote mathematics textbooks, teachers' and workbooks, and supplementary resource books, which are approved by the Ministry of National Education and used as an official textbook in the country. Based on the experience of teaching she is interested in student-centered instructional design, reflective practice, design-oriented and interdisciplinary learning, differentiated, and data-driven instructional design.

María Zepeda Hurtado is a researcher professor at the Instituto Politécnico Nacional, CECyT 11, México, with thirty-five years of teaching experience, instructor of courses, workshops, seminars, designer of teacher training actions in the pedagogical area.

Index

R

S

T

U

www.ingramcontent.com/pod-product-compliance
Lightning Source LLC
LaVergne TN
LVHW080040210926
848481LV00011B/3243